Lecture Notes in Computer Science 4824

Commenced Publication in 1973
Founding and Former Series Editors:
Gerhard Goos, Juris Hartmanis, and Jan van Leeuwen

Adrian Paschke   Yevgen Biletskiy (Eds.)

Advances in Rule Interchange and Applications

International Symposium, RuleML 2007
Orlando, Florida, October 25-26, 2007
Proceedings

 Springer

Volume Editors

Adrian Paschke
Technical University Dresden
Biotec/Dept. of Computing
Tatzberg 47-51, 01307 Dresden, Germany
E-mail: adrian.paschke@biotec.tu-dresden.de

Yevgen Biletskiy
University of New Brunswick
15 Dineen Drive, Fredericton, New Brunswick, E3B 5A3, Canada
E-mail: biletski@unb.ca

Library of Congress Control Number: 2007937616

CR Subject Classification (1998): D.3.1, F.3.2, H.5.3

LNCS Sublibrary: SL 2 – Programming and Software Engineering

ISSN   0302-9743
ISBN-10   3-540-75974-3 Springer Berlin Heidelberg New York
ISBN-13   978-3-540-75974-4 Springer Berlin Heidelberg New York

Springer is a part of Springer Science+Business Media

springer.com

© Springer-Verlag Berlin Heidelberg 2007
Printed in Germany

Typesetting: Camera-ready by author, data conversion by Scientific Publishing Services, Chennai, India
Printed on acid-free paper   SPIN: 12180622   06/3180   5 4 3 2 1 0

Preface

The International Symposium on Rule Interchange and Applications (RuleML-2007), collocated in Orlando, Florida, with the Tenth International Business Rules Forum, was the first symposium devoted to work on practical distributed rule technologies and rule-based applications which need language standards for rules operating in the context of modern infrastructures, including the Semantic Web, intelligent multi-agent systems, event-driven architectures, and service-oriented computing applications. The symposium was organized by the RuleML Initiative, financially and technically supported by industrial companies (Top Logic, VIStology, and Inferware) and in cooperation with professional societies (ECCAI, AAAI, ACM, ACM SIGAPP, ACM SIGMIS, ACM SIGART, ACM SIGMOD, IEEE, IEEE Computer TCAAS, IEEE SMCS, BPM-Forum, W3C, OMG, and OASIS).

The RuleML Initiative is organized by representatives from academia, industry and government for the advancement of rule technology, providing enhanced usability, scalability and performance. The goal of RuleML (www.ruleml.org) is to develop an open, general, XML-based family of rule languages as intermediaries between various 'specialized' rule vendors, applications, industrial and academic research groups, as well as standardization efforts such as OMG's PRR or W3C's RIF. A general advantage of using declarative rules is that they can be easily represented in a machine-readable and platform-independent manner, often governed by an XML schema. This fits well into today's distributed, heterogeneous Web-based system environments. Rules represented in standardized Web formats can be discovered, interchanged and invoked at runtime within and across Web systems, and can be interpreted and executed on any platform.

After a series of successful RuleML workshops and then conferences (e.g., http://2006.ruleml.org), the RuleML 2007 Symposium was a new kind of event where the Web Logic community joined the established, practically oriented Forum of the Business Rules community (www.businessrulesforum.com) to help the cross-fertilization between Web and business logic technology. The goal of RuleML 2007 was to bring together rule system providers, representatives of, and participants in, rule standardization efforts (e.g., RuleML, RIF, SBVR, etc.) and open source rule communities (e.g., jBoss Rules, Jess, Prova, OO jDREW, etc.), practitioners and technical experts, developers, users, and researchers, to attend an exciting venue and exchange new ideas, practical developments and experiences on issues related to the engineering, management, integration, interoperation, and interchange of rules in open distributed environments such as the Web.

In spite of this year's late announcement, RuleML 2007 received 41 original submissions from contributors around the world. The organizers of this event could take advantage of the very strong International Program Committee, who rigorously reviewed the submissions. Each submission was reviewed at least by three Program Committee members. Finally, the symposium program included: nine full papers, which presented solid work in advancing and assessing the state of the art in rule-based systems, including event processing systems; two invited papers accompanying keynotes; nine short papers, which presented innovative ideas, approaches, and implementations; and three demo papers accompanying presentations of practical applications at the RuleML 2007 Challenge session. Full papers represented two main streams: (1) Rule languages, rule engines, and rule interchange. (2) Business process and service modeling and management. Short papers covered several aspects including advances in rule languages, rule engines, integration of rules and ontologies, as well as rule applications.

To place emphasis on the practical use of rule technologies in distributed Web-based environments, there was a RuleML 2007 Challenge with a focus on rule engines and rule technologies, interoperation, and interchange. The challenge offered participants the opportunity to demonstrate their commercial and open source tools, use cases, and applications.

Further highlights of RuleML 2007 included:

- Plenary keynotes given by world-class practitioners and researchers, featuring foundational topics of rule-based computing and industrial success stories:
- 1. Ron Ross (BRCommunity):
   "Business Requirements for Rule Modeling"
   2. Jürgen Angele (Ontoprise):
   "How Ontologies and Rules Help to Advance Automobile Development"
   3. Adrian Bowles (OMG):
   "Sharing Policy Rules for IT Governance"
   4. Harold Boley (NRC IIT):
   "Are Your Rules Online? Four Web Rule Essentials"
- A panel discussion on Web business rules, featuring prominent and visionary panelists.
- Social events to promote networking among the symposium delegates.

The RuleML 2007 organizers wish to thank the excellent Program Committee for their hard work in reviewing the submitted papers. Their criticism and very useful comments and suggestions were instrumental in achieving a high-quality publication. We also thank the symposium authors for submitting good papers, responding to the reviewers' comments, and abiding by our production schedule. We further wish to thank the keynote speakers for contributing their interesting talks. We are very grateful to the Business Rules Forum organizers for enabling this fruitful collocation of the Tenth Business Rules Forum and RuleML 2007. In particular, we thank Gladys Lam and Valentina Tang for their support. Finally, we thank our sponsors, whose financial support helped us to offer this event, and whose technical support allowed us to attract many high-quality submissions.

August 2007   Adrian Paschke
Yevgen Biletskiy

Organization

Program Committee

Asaf Adi, IBM Research Laboratory Haifa, Israel
Bill Andersen, Ontology Works, USA
Grigoris Antoniou, University of Crete, Greece
Arun Ayachitula, IBM T.J. Watson Research Center, USA
Youcef Baghdadi, Sultan Qaboos University, Oman
Sidney Bailin, Knowledge Evolution, USA
Claudio Bartolini, HP Labs Palo Alto, USA
Tim Bass, SilkRoad Inc., USA
Nick Bassiliades, Aristotle University of Thessaloniki, Greece
Bernhard Bauer, University of Augsburg, Germany
Lepoldo Bertossi, Carleton University, Canada
Anthony Brown, University of New Brunswick, Canada
Loreto Bravo, University of Edinburgh, UK
Diego Calvanese, Free University of Bozen-Bolzano, Italy
Donald Chapin, Business Semantics Ltd., UK
Keith Clark, Imperial College, UK
Jorge Cuellar, Siemens AG, Germany
Mike Dean, BBN Technologies, USA
Stan Devitt, Agfa Healthcare, Canada
Jens Dietrich, Massey University, New Zealand
Jürgen Dix, Technische Universität Clausthal, Germany
Scharam Dustdar, Vienna University of Technology, Austria
Andreas Eberhart, HP, Germany
Maarten van Emden, University of Victoria, Canada
Opher Etzion, IBM Research Laboratory Haifa, Israel
Dieter Fensel, DERI Innsbruck, Austria
Tim Finin, University of Maryland Baltimore County, USA
Dragan Gasevic, Athabasca University, Canada
Allen Ginsberg, MITRE, USA
Robert Golan, DBmind Technologies, USA
Guido Governatori, University of Queensland, Australia
Sung Ho Ha, Kyungpook National University, South Korea
Gary Hallmark, Oracle, USA
Marek Hatala, Simon Fraser University, Canada
Ivan Herman, W3C Semantic Web Activity, Netherlands
David Hirtle, University of Waterloo, Canada
Jochen Hiller, Top Logic, Germany
Christian Huemer, Vienna University of Technology, Austria
Jane Hunter, University of Queensland, Australia

Table of Contents

Session: RuleML-2007 Challenge

Session: Rules, Reasoning, and Ontologies

Session: Reaction Rules and Rule Applications

How Ontologies and Rules Help to Advance Automobile Development

Thomas Syldatke[1], Willy Chen[2], Jürgen Angele[3], Andreas Nierlich[3], and Mike Ullrich[3]

[1] Audi AG, Ingolstadt
Thomas.syldatke@audi.de
[2] Achievo Inproware GmbH, Ingolstadt
willy.chen@achievo.com
[3] ontoprise GmbH, Karlsruhe
{angele,nierlich,mike.ullrich}@ontoprise.de

Abstract. Nowadays the increasing complexity of cars has become a major challenge due to the growing rate of electronic components and software. This trend has an impact on all phases of the life cycle of the car. Especially during the process of testing cars and components semantic based approaches can help to increase the automation ratio of testing components and to find the needle in the haystack. Ontologies in combination with rules have proven to check specifications against test results and therefore deliver new insights.

Keywords: ontology, rule, automotive engineering.

1 Introduction

A multiplicity of today's innovations in the automotive industry is reached by new and more efficient electronic systems (e.g. Multimedia, drivers' assistance systems, safety systems, etc.). Especially for premium class cars the number of integrated controllers increases constantly. Therefore not only the development of such systems but also their integration into the car is accordingly complex. The essential reduction of time-to-market and thus shortened development cycles is even enhancing the complexity and intensifying the problem.

To be able to cope with these requirements car manufacturers as well as their suppliers have to implement highly efficient test methods for the electronic systems.

Testing is an essential part in the development of a new car to ensure an outstanding quality of the final product. This is also true for embedded software and electronic components. While such systems have been quite simple in the past, they are nowadays getting more and more complex. The complexity is thereby not only driven by the actual functionality but also by the application in different car configurations and environments. Furthermore, their number has grown up to 60 electronic control units (ECU) in a single premium class car.

An important task in the overall testing process is the simulation of a physical ECU in a virtual car environment, which is also known as Hardware–in–the–Loop (HiL) test. Unlike tests in real prototypes, critical situations and system states can be covered without any risk. Targets during HiL tests are the validation of the ECU

A. Paschke and Y. Biletskiy (Eds.): RuleML 2007, LNCS 4824, pp. 1–6, 2007.

against its specification and other given requirements. The main problem thereby is the huge amount of data, which is being generated and recorded during such a test run. Therefore, the analysis of that data is not only very time-consuming, but one could easily overlook a fault.

Existing software solutions do assist the engineers in performing the analysis by providing different visualizations of the recorded data. However, they lack a standardized way to define the desired system behaviour on the one hand and known error cases on the other. The application of ontologies to provide a uniform vocabulary for that domain and rules to specify the system behaviour as well as error cases have shown promising results to face these issues and automate this part of the testing process.

This paper will give a brief survey over a prototype realized for Audi. Section 2 summarizes the concrete business use case. The prototype itself is presented in section 3 and discussed in section 4. Finally, section 5 concludes the paper and gives some plans and ideas for the future work on this topic.

2   Use Case

In 2006 Audi has announced the first engines including the Audi valvelift system. "This innovation varies the valve lift between two levels. To achieve this, sets of sliding cams are mounted directly on the intake camshafts. These feature two sets of adjacent cam contours for small and large valve lift. Which cam is used to open the intake valves depends on the power demand at any one time. The effect is an appreciable increase in engine efficiency. The driver benefits from greater power and improved driveability, while enjoying a marked reduction in fuel consumption"[1].

The valve lifts are controlled by the engine management system. A deterministic finite automaton3 internally represents the possible states of the valvelift system and specifies the conditions for switching the valve lifts. This automaton has 6 different states $S_1,..,S_6$ whereas S_1 is the start state. S_1 and S_4 represent the states in which the small respectively the large valve lift is being used. The other ones are transition states that have to be passed in a given order. The state transition functions δ finally define when to switch between the valve lifts. An example for such a transition function is: "If the engine speed is greater than 4000 the valvelift system must switch to S_4 if it is in S_1". These details are defined in the specification of the engine management system.

Without knowing any internal details about the engine management system an engineer could also specify the expected behaviour of the valvelift system from another point of view - e.g.: "at idle speed the small valve lift must be used". During expert interviews we have been able to elicit 18 of these kind of rules that need to be kept in mind when testing the engine management system.

3   The Prototype

The main challenge for a first prototype is to merge both sources of knowledge about the valvelift system and provide a way to automatically validate and analyze the data recorded during the HiL tests. Therefore, a flexible way of formalizing the given rules for the transition functions as well as the system constraints is required.

Adopting ontologies for defining a uniform domain vocabulary and rules is a promising way to meet these demands. Both have recently gained increasing attention from researchers as well as the industry, because they are considered key technologies for enabling the vision of a Semantic Web. First companies are also starting to provide professional support in that technology niche.

Since researchers still disagree about the best way of combining (OWL) ontologies and rules, we have selected the representation language that comes with the best professional support. F–Logic [2] together with the inference machine OntoBroker [3] from Ontoprise turned out to be most suitable for our demands.

3.1 Domain Ontology

The first task in the prototype development is the modelling of a domain ontology that provides a common vocabulary for the definition of rules. We thereby need to take the format of the recorded data from HiL tests into account, since they will in the end be the facts of our domain ontology.

Using the tool INCA it is possible to take snapshots of internal variables of an ECU (e.g., current state of the finite automaton, engine speed, oil temperature) during the HiL tests with a minimum time interval of approx. 10ms. Table 1 gives an example of such a recorded data set.

Table 1. Example data set from INCA

time(ms)	var_1	var_2	var_3	var_4	..	var_{24}
0.003264454	2519	1	2 90	..	15005	
0.013361665	2519	1	2 90	..	15006	
..	..	..	..	..	..	..

To keep the first prototype quite simple, we define two concepts:

- *Situation*: represents a single snapshot of the ECU at a given time including the recorded variables and a relation to the successor snapshot if present.
- *State*: the facts of this class represent the possible states of the finite automaton.

3.2 System and Expert Rules

The basic idea for defining rules on top of the domain ontology is on the one hand to simulate the expected state that should be present at a given snapshot. Therefore we add a relation *nextState* to the concept *Situation* and can derive its value based on the given transition functions. The above mentioned rule ("If the engine speed is greater than 4000 the valvelift system must switch to S4 if it is in S1") is for example represented in F–Logic as follows:

```
?S[nextState->S4] <-
?S:Situation[state->S1,engineSpeed->?V] and ?V > 4000.
```

On the other hand we can use this derivation results to check, whether it is equal to the measured state. In F-Logic this is represented as a constraint:

```
! ?S:Situation[nextState->?X,successor->?SS] and not
?SS[state->?X].
```

Based on this approach we can seamlessly integrated the expert rules by formulating them as error statements:

```
! ?S:Situation[state->S4,IDLE->>1].
```

The OntoBroker offers the possibility to retrieve the proof tree for a specific query result – i.e. which variables and values have been used in which rule. This feature can be utilized to define explanations for rule. A detailed description can be found in [3].

3.3 The Prototype System

Figure 1 shows the prototype system, which is embedded as a plugin in an Eclipse RCP-based applications, which also includes a module to manage ontologies and rules. The main features are:

- Import HiL test data: The data recorded with INCA can be transformed into instances of the domain ontology.
- Simulation Run: By applying the defined rules, the test data instances are checked for possible errors, which are then highlighted.
- Explanation of results: If available, an explanation for a detected error is presented. Tests with real data from HiL tests have shown reasonable results. Furthermore, some minor errors could be detected as well as explained. Even with a large number of instances (approx. 50.000) the prototype performed very well.

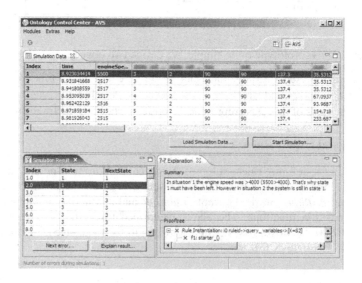

Fig. 1. Screenshot of the prototype

4 Discussion of the Prototype

The prototype system has been accepted very positively by the responsible engineers and do address their current needs. We have also been able to show the value of the application of ontologies and rules. The most important benefits are thereby:

- Integration of specification and expert knowledge: With ontology and rules, it has been possible to merge two completely different sources of knowledge and make them processable by computers. Therefore, a high automation of performing data analysis of HiL test results could be achieved.
- Evolving rule bases: The rule base does not necessarily need to be complete at any time. In fact the better the rules are the more errors could be detected automatically. Changing a rule can also be done very quickly. It is important to stress that the presented prototype does not aim to replace current testing technologies or experts but to support the engineers in doing their job more efficiently.
- Sharing and reuse of knowledge: Rules that capture specific knowledge can be centrally stored and made available to a broader audience. Knowledge islands could thus be avoided. In addition, each data analysis is able to apply all previously defined rules. Thus the knowledge about why an error has occurred is not getting lost.
- Explanation of results: The usage of appropriate inference machines allows a detailed explanation of the used rules with only a minimal amount additional work. This may give an engineer a clue for finding and solving an error.

5 Conclusion and Outlook

The complexity of requirements specifications of car components is steadily growing due to the increasing number of software and electronic components in cars and due to the dependency of these components to each other. Each additional electronic component may add up to exponential growth in complexity of dependencies. This results in impossibility for humans to track all dependencies. State of the art algorithms lack the ability of formalizing the underlying knowledge in both, a human and a machine processable way – thus leading to a reduced analysis during the test process. Additionally the formalized knowledge has to be connected with physical test results in a scalable manner.

To overcome these obstacles an ontology and rule based approach on basis of F-Logic has been chosen. F-Logic combines

- a formalization mechanism close to technical language,
- powerful modeling primitives, strong enough to express the dependencies and complexities of technical specification,
- the ability to describe technical formulas and functions,
- an integration mechanism, even capable of ad-hoc integration of current test results,
- and existing paradigms such as deductive databases, logic programming, ontologies and descriptive rules.

With the use of F-Logic the existing knowledge can be formulated in a separate layer, aside the application and aside the data vaults.

The application of ontologies and rules for supporting the testing process of ECUs has shown promising results. We have successfully convinced the decision makers and are now working on the integration of additional ECU functions. Future plans and ideas for evolving the presented approach are:

- Support of different vendors: Various electronic control units are offered by different vendors with exactly the same functionality but different internals. Thus, an adequate support by layered vendor ontologies would be desirable.
- Support regulatory requirements: Every market (e.g. European Union, US) has different regulatory requirements. Instead of defining rules for each market, the regulations could be modelled in market ontologies. Depending on the market an ECU should be tested for, the appropriate ontology can be applied dynamically during data analysis.
- Automated extraction of rules: Since technical specifications are highly formalized, it could be possible to automate the extraction of the essential rules and therefore provide a seamless integration into today's testing infrastructure.

We have shown that ontologies may very well be used to enhance business processes in our special case testing of electronic control units in cars. It has been shown that complex dependencies as well as compliance rules are very well represented using ontologies represented in F-Logic. Especially rules play a crucial role in this application.

This application accelerates the testing process and increases quality at our customer and thus enhances and accelerates the development of new cars as well. This reduces time-to-market in the very end.

Acknowledgement

Achievo Inproware GmbH has been responsible for the development of the presented prototype. Their work has been supported by Ontoprise. Both companies are now cooperating as partners to bring that approach to a broader audience.

References

[1] Audi. The new Audi A5 / Audi S5 (2007), http://tinyurl.com/3bv6u3
[2] Kifer, M., Lausen, G., Wu, J.: Logical foundations of object-oriented and frame-basedlanguages. JACM: Journal of the ACM 42, 741–843 (1995)
[3] Ontoprise. How to Use OntoBroker - Users and Developers Guide for the OntoBroker - System / Version 4.3 (2006)

Are Your Rules Online?
Four Web Rule Essentials

Harold Boley

Institute for Information Technology – e-Business
National Research Council of Canada
Fredericton, NB, E3B 9W4, Canada
`harold.boley@nrc.gc.ca`

Abstract. Four principal Web rule issues constitute our starting points: I1) Formal knowledge representation can act as content in its own right and/or as metadata for content. I2) Knowledge on the open Web is typically inconsistent but closed 'intranet' reasoning can exploit local consistency. I3) Scalability of reasoning calls for representation layering on top of quite inexpressive languages. I4) Rule representation should stay compatible with relevant Web standards.

To address these, four corresponding essentials of Web rules have emerged: E1) Combine formal and informal knowledge in a Rule Wiki, where the formal parts can be taken as code (or as metadata) for the informal parts, and the informal parts as documentation (or as content) for the formal parts. This can be supported by tools for Controlled Natural Language: mapping a subset of, e.g., English into rules and back. E2) Represent distributed knowledge via a module construct, supporting local consistency, permitting scoped negation as failure, and reducing the search space of scoped queries. Modules are embedded into an 'Entails' element: prove whether a query is entailed by a module. E3) Develop a dual layering of assertional and terminological knowledge as well as their blends. To permit the specification of terminologies independent of assertions, the CARIN principle is adopted: a terminological predicate is not permitted in the head of a rule. E4) Differentiate the Web notion of URIs as in URLs, for access, vs. URNs, for naming. A URI can then be used: URL-like, for module import, where it is an error if dereferencing does not yield a valid knowledge document; URN-like, as an identifier, where dereferencing is not intended; or, as a name whose dereferencing can access its (partial) definition.

Introduction

Web rules constitute an effort towards novel Web sites with machine-interpretable rule representations for automated reasoning.

This research builds on our previous work in Web rule foundations (e.g., POSL[1], Datalog^{DL} [MLB+07b], ALC^u_P [MLB07a]), standards (e.g., RuleML

---

[1] http://www.ruleml.org/submission/ruleml-shortation.html

A. Paschke and Y. Biletskiy (Eds.): RuleML 2007, LNCS 4824, pp. 7–24, 2007.

[Bol06], SWRL [HPSB+04], RIF [BK07]), engines (e.g., OO jDREW [BBH+05]), as well as use cases (e.g., AgentMatcher [BBY04], NBBizKB [MB05], FindXpRT [LBBM06], Rule Responder [PKB07], and Ontology Integration [BHV06]). Its objective is to devise complementary rule representation and reasoning techniques for Business Rules, the Semantic Web, Web Services, as well as other Web (and Web 2.0) areas.

Previous research led to the following four principal Web rule issues used here as starting points.

I1. The Web increasingly develops 'Semantic Subwebs' containing knowledge documents (knowledge bases, schemas, etc.). A formal knowledge representation can act as content that is queried and retrieved in its own right, as metadata that helps to retrieve other formal or informal content, or as a combination of both [Bol99].

I2. The open Web as well as some closed 'intranets' contain knowledge documents. Open Web knowledge (knowledge distributed over an open-ended set of Web documents) in expressively rich representations is typically inconsistent whereas closed intranet knowledge (knowledge maintained in closed sets of Web documents) is typically *paraconsistent*, i.e. each intranet is kept (locally) consistent, although their union may be (globally) inconsistent. Thus, while open Web reasoning cannot directly use classical 2-valued logic, closed intranet reasoning can exploit it locally. Beyond benefiting from an implicit module notion through documents, locality can be achieved via an explicit module construct.

I3. Besides the consistency issue I2, the trade-offs between representation expressiveness and reasoning tractability imply that scalability of reasoning to the open Web is still unresolved for higher expressive classes. This calls for representation layering on top of quite inexpressive languages. W3C's Semantic Web stacks, old and new [KdBBF05], have made a start, and introduced RDF as the fundamental knowledge layer. However, the (official) XML syntax of RDF is somewhat complicated (although its triple syntax is extremely simple and its N3-like syntaxes are quite popular), and the (official) semantics of RDF [Hay04] is rather complicated [Sch05]. Yet, simple RDF statements without blank nodes (existentials) in assertions, queried without property variables (existentials for predicates), are a candidate for the least expressive (binary-)fact layer. Binary Datalog rules, similar to relational views over 2-column tables, can then be added to derive new facts from conjunctions of other facts, much like relational joins. Finally, an irreflexive `subClassOf` fragment of RDF Schema can be employed to define order-sorted types for constants and rule variables.

I4. Since Web rules are layered on, and side-by-side with, existing Web languages, it is important to represent rules so that compatibility with relevant Web standards (e.g., XML, RDF, OWL) is preserved. The selection of Web standards can be hard, for instance when considering which, if any, query and transformation languages should be included (e.g., XQuery, XSLT, SPARQL, OWL-QL)[2]. Likewise, the (syntactic and semantic) levels and degrees of

---

[2] http://www.dajobe.org/talks/200603-sparql-stanford

compatibility with each of the selected languages need to be determined. Therefore, essential E4 will focus on what is unique to Web languages, namely 'webizing' [BL01], basically permitting the Web rule language to use URIs for global constants, in ways compatible with URIs in existing (Semantic) Web languages.

To address these issues, the following sections will consider four corresponding essentials of Web rules. For a tutorial-style survey on Web rules see [BKPP07].

E1 Combining Logic Rules with Controlled English

We propose to combine formal and informal knowledge in a **Rule Wiki**, where *clauses* (rules and, as a special case, facts) are given dual representations, in a natural language and in a logic language. The formal parts can be taken as code (or as metadata) for the informal parts, and the informal parts as documentation (or as content) for the formal parts. On the level of a single rule, both representations can be kept separate (our assumption in the following) or can be intertwined. This combination is analogous to Literate Programming [Knu84] and Javadoc. It may be supported by tools mapping Controlled Natural Languages (e.g., Controlled English) into rules and back. For instance, two English-to-rule tools based on Attempto [FHK+05] are TRANSLATOR [Hir06] and AceRules [Kuh07]. Related tools have also been developed for AceWiki[3] and "Semantics of Business Vocabulary and Business Rules" (SBVR) [BHC05].

A classical Wiki permits the authoring of informal-knowledge documents using natural-language-enriching markup simpler than HTML. Extending this concept, a Rule Wiki permits formal-knowledge authoring using logic-language-enriching markup simpler than HTML or XML, combining this with informal-knowledge authoring. The formal-knowledge language can employ a human-readable syntax such as POSL[1], integrating the Prolog and F-logic syntaxes.

Let us consider an example. In a logistics use case, the ternary relation `reciship` can represent reciprocal shippings of unspecified cargos at a total cost between two sites. A Datalog rule infers this conclusion from three premises, two `shipment` atoms and an `add` atom. The `shipment` relation is based on slotted facts, and the `add` relation, on a SWRL built-in. Below is a Controlled English representation using MediaWiki[4] markup, where commutative premise conjuncts, the default in POSL, become "*"-unordered list bullets ("#"-ordered lists could be used for sequential conjunctions), predicate/built-in and slot names (both of which could be hyperlinked to their definitions, cf. section E4) are bold-faced with triple apostrophes, and variable names are italicized with double apostrophes; the logical connectives if, and, etc. are marked up with double brackets as internal links (possibly redirecting via interwiki links) to their explanatory pages. For the proposed **Rule MediaWiki**, this is followed by a POSL representation, which will also be employed and developed in subsequent sections. The POSL rule markup exemplifies the use of Prolog's ":-" infix between

---

[3] http://attempto.ifi.uzh.ch/acewiki
[4] http://en.wikipedia.org/wiki/MediaWiki

the conclusion and premises, top-level commas for separating conjuncts, intra-atom commas for ordered arguments and semicolons for unordered arguments, F-logic's infix "->" for slots, as well as the prefix "?" for named variables and a stand-alone "?" for anonymous variables:

```
A reciprocal shipping, '''reciship''', at total amount, ''cost'',
takes place between sites ''A'' and ''B'' [[if]]
* a '''shipment''' has a '''cargo''', a '''price''' of ''cost1'', a
'''source''' of ''A'', and a destination, '''dest''', of ''B'' [[and]]
* a '''shipment''' has a '''cargo''', a '''price''' of ''cost2'', a
'''source''' of ''B'', and a destination, '''dest''', of ''A'' [[and]]
* ''cost'' is the addition, '''add''', of ''cost1'' and ''cost2''.
```

```
 reciship(?cost,?A,?B) :-
    shipment(cargo->?;price->?cost1;source->?A;dest->?B),
    shipment(cargo->?;price->?cost2;source->?B;dest->?A),
    add(?cost,?cost1,?cost2).
```

Fig. 1 shows the actual rendering of the MediaWiki part, followed by the proposed rendering of its POSL extension, employing corresponding fonts in Controlled English and POSL.

A reciprocal shipping, **reciship**, at total amount, *cost*, takes place between sites *A* and *B* <u>if</u>
- a **shipment** has a **cargo**, a **price** of *cost1*, a **source** of *A*, and a destination, **dest**, of *B* <u>and</u>
- a **shipment** has a **cargo**, a **price** of *cost2*, a **source** of *B*, and a destination, **dest**, of *A* <u>and</u>
- *cost* is the addition, **add**, of *cost1* and *cost2*.

```
reciship(?cost, ?A, ?B) :-
    shipment(cargo->?;price->?cost1;source->?A;dest->?B),
    shipment(cargo->?;price->?cost2;source->?B;dest->?A),
    add(?cost, ?cost1, ?cost2).
```

Fig. 1. A Rule Wiki example

For Web interchange, POSL can be automatically converted to RuleML/XML, and vice versa, using our online translator[5] via Java Web Start. The result of serializing the POSL example in RuleML/XML is shown in appendix A.1.

E2   A Distributed Rule Module Construct

It is beneficial to represent distributed knowledge via a module construct, supporting local consistency, reducing the search space of scoped (module-restricted) queries and permitting scoped negation as failure (over closed worlds).

Such Web modules may be written and used 'in place' or defined at one place (URL) and accessed from other places. The semantics of modules should not depend on any needed URL-dereferencing.

---

[5] http://www.ruleml.org/posl/converter.jnlp

The granularity of modules may vary: (1) It is quite possible to specify multiple modules within one knowledge document, even allowing different modules specified in the premise conjuncts of a single rule. (2) It is very natural to specify one module per (URL-accessed) document. (3) It is somewhat unusual to specify multiple documents constituting one module.

Versions of contexts/modules have been studied in AI [McC93] and Prolog [HF06], and are used in the Semantic Web as TRIPLE models [SD02], F-logic's scoped formulas [YK03], N3 formulae [BL06], and RDF named graphs [SvESH07]. In RuleML and its OO jDREW implementation, modules take the form of a `Rulebase` element, which is assumed within a top-level `Assert`.

Initially, a flat set of modules can be studied on the basis of McCarthy's `ist` operator to prove whether a query is true in a module. In recent RuleML releases, modules have been embedded into an `Entails` element, which serves to prove whether a query or module is entailed by another module.

This can then be extended to a nested (cycle-free) inheritance system of modules, as surveyed in [BLM94]. We only need to consider a simplified kind of module inheritance here, since by default we do not assume a Prolog-like textual order in a module's set of assertional (fact and rule) or terminological (subclass-ontological) clauses and therefore do not need to merge clauses but can just take their union.

Another simplification occurs if a module consists only of clauses for a single predicate (as in Prolog dubbed a *procedure*). For such a module, an explicit syntactic grouping element might be introduced, intermediate between the levels of a clause and of a general module. The granularity of procedures relative to general modules may vary similarly to the above granularity of general modules relative to documents.

As an example let us consider two modules, which could reside in a single document or (as serialized in appendix A.2) each constitute a separate document. Both contain rules granting or denying discounts, which could be extended with rules about giveaways etc. The first, of the special form of a `discount` procedure, might be called the `loyalty` module, as it grants discount percentages to loyal customers for certain products (module contents will be enclosed using "{...}"):

```
{
    discount(?customer,?product,percent[5]) :-
        premium(?customer),
        regular(?product).

    discount(?customer,?product,percent[10]) :-
        premium(?customer),
        luxury(?product).
}
```

Note that the percent sign and units of measurement are written as (uninterpreted) unary functions applied to a value. For such 'passive' functions, the arguments will be enclosed using "[...]", where the entire application forms an expression that acts as a complex term (in Prolog called a 'structure').

The second might be called the `legality` module, as it denies discounts to customers who used fraudulent payment methods or who payed more than 45 days after a delivery (a "−" prefix is used for the strong negation of atoms):

```
{
  -discount(?customer,?product1,?percent) :-
    payment(?customer,?product2,?amount,?method,?time),
    fraudulent(?customer,?method,?time).

  -discount(?customer,?product1,?percent) :-
    delivery(?customer,?product2,?amount,date[?y1,?m1,?d1]),
    payment(?customer,?product2,?amount,?method,date[?y2,?m2,?d2]),
    datediff(days[?delta],date[?y2,?m2,?d2],date[?y1,?m1,?d1]),
    greaterThan(?delta,45).
}
```

Note that dates are expressed here as complex terms of three arguments: the year, followed by the month, followed by the day.

These modules are locally consistent, although their union is inconsistent. For instance, according to the first rule of the `loyalty` module, premium customers would be granted 5 percent discount for a regular product, but, according to the first rule of the `legality` module, would be denied any discount on any product if they used a fraudulent payment method on a product. To deal with this, *prioritization* (cf. Courteous Logic Programs [Gro04] and Defeasible Logic [BAV04]) could be employed on the module level to let all rules of the `legality` module *override* all `loyalty` rules.

Using `Naf` as the primitive for negation-as-failure, and "|−" as the infix for the `Entails` element, a negated-as-failure *query* scoped by a *module* can be achieved via *module* |− Naf(*query*). E.g., `loyalty` |− Naf(giveaway(John,Mercedes)) is true, since the `loyalty` module does not entail that John obtains a Mercedes giveaway (or any other giveaway).

Scoping is also helpful for positive queries, since the search space can be reduced if the modules to be searched are known beforehand. For example, in the `loyalty` module, the queries in the premises of the `discount` rules can be sped up by restricting them to `customer` and `product` modules:

```
{
  discount(?customer,?product,percent[5]) :-
    customer |- premium(?customer),
    product |- regular(?product).

  discount(?customer,?product,percent[10]) :-
    customer |- premium(?customer),
    product |- luxury(?product).
}
```

Moreover, to check whether a rulebase *KB* obeys integrity constraints in *IC*, we use the approach of [Rei88] to check whether module *KB* entails module *IC*.

E3 Assertional-Terminological Expressiveness Layering

Quite an effort has been made to develop a dual expressiveness layering of assertional and terminological knowledge as well as their blends [ADG+05, KdBBF05].

To retain decidability of querying, the *assertional bottom layer* usually consists of Datalog (function-free) assertions, perhaps restricted to unary/binary predicates. For the *terminological bottom layer*, an irreflexive version of RDF Schema's subClassOf can be employed, which could later be extended towards the ρDF [MPG07] fragment of RDF. The two layers can be blended through a hybrid combination (ρDF classes used as types for Datalog constants and variables, and subClassOf defined with order-sorted semantics) or a homogeneous integration (ρDF classes used as unary predicates in the body of Datalog rules, and subClassOf defined as special rules with Herbrand-model semantics).

The higher layers can develop Datalog into Horn (as in OO jDREW's hybrid implementation) and FOL (First-Order Logic) assertions, ρDF into ALC and SHIQ terminologies with classes and properties, and appropriate blends [Ros06], e.g. as advancements of our hybrid Datalog^{DL} [MLB+07b] or homogeneous ALC^u_P [MLB07a]. For certain purposes, especially in the early modeling phases, the assertional layers can move even beyond FOL, including towards higher-order and modal logics, as started as part of the RuleML family [Bol06].

To permit the specification of terminologies independent of assertions, a hybrid approach is proposed here adopting the CARIN [LR98] principle as a working hypothesis: A terminological predicate is not permitted in the head of a rule. Intuitively, terminological classes cannot be (re)defined by assertional clauses, because a terminology establishes more stable 'background' knowledge extended by assertions that constitute more volatile 'foreground' knowledge.

Such a hybrid lower layer can use sort restrictions as simple terminological queries in Datalog rule bodies, which in higher layers are extended to terminological queries involving properties, ALC expressions, etc. In the spirit of [KdBBF05], this should lead to a more realistic Semantic Web architecture with simplified foundations and better computational properties. Our fine-grained bottom-up approach also complements the recent differentiation of OWL-Lite into OWL 1.1 Tractable Fragments [GCG+06].

The following example uses classes of a subClassOf terminology as variable sorts of slightly extended Datalog rules, namely of Horn logic rules employing (unary) functions only for measurement units.

The terminology forms a DAG that introduces Vehicle-rooted classes and exemplifies multiple inheritance of MiniVan from Van and PassengerVehicle (the ">" infix is used between a superclass and a subclass):

```
Vehicle > Van
Vehicle > PassengerVehicle
Van > MiniVan
PassengerVehicle > MiniVan
PassengerVehicle > Car
```

The rules (serialized in appendix A.3) specify registration fees for vehicles. The first rule specifies a vehicle variable sorted by the Van class, while the second refers to the Car class (the ":" infix is used between a variable and its sort):

```
registration(?V:Van,CAD[?R:Decimal]) :-
  emission(?V,CO2[?E]),
  weight(?V,kg[?W]),
  emiweight(CAD[?R],CO2[?E],kg[?W]).

registration(?V:Car,CAD[?R:Decimal]) :-
  emission(?V,CO2[?E]),
  speed(?V,kmh[?S]),
  emispeed(CAD[?R],CO2[?E],kmh[?S]).
```

A registration query for a given vehicle class will thus unify only with correspondingly sorted rule conclusions, hence directly branch into the appropriate rule premises (the emiweight and emispeed premises compute the fees from the emissions as well as the weights and speeds for Vans and Cars, respectively). Section E4 will show URI-'webized' versions of these terminological classes.

E4   URIs for Access, Naming, or Both

There have been attempts to differentiate the Web notion of URIs into two sub-notions, as discussed in [Hal06]: URLs (Uniform Resource Locators), for access, and URNs (Uniform Resource Names), for naming. This distinction is independent from the recent IRI (Internationalized) versions of URIs. In the context of Web knowledge representation, especially for Web rules as explored in POSL, RuleML, and RIF, three central URI uses are emerging, given here in the order of further needed research (orthogonal to research in URI normalization [Bol03]).

First, a URI can be used, URL/access-style, for module import (transitive import for nested modules), where it is an error if dereferencing the URI does not yield a knowledge base valid with respect to the expected representation language.

Second, a URI can be used, URN/naming-style, as the identifier of an individual constant in the representation language, where URI dereferencing is not intended as part of the formal knowledge representation. If dereferencing is attempted as part of the metadata about the informal knowledge representation, it should retrieve a 'homepage' of the individual; cf. section E1.

Third, a URI can be used, naming-style, as the identifier of a class, property, relation, or function, and at the same time, access-style, where dereferencing yields (a "#"-anchor into) a knowledge base formally defining that identifier (albeit perhaps partially only, as for an RDF Schema knowledge base just giving the superclasses of a class).

Here are examples for the three URI uses in connection with rules.

First, the `loyalty` module of section E2 could be imported into the current rulebase using the URL/access-style URI `http://modeg.org#loyalty`.

Second, the URI `http://en.wikipedia.org/wiki/Pluto` can be used URN/naming-style to refer to a celestial body originally considered a planet, as in this rule (serialized in appendix A.4) specifying its years of planethood (a URI is enclosed in a pair of angular brackets, `<...>`):

```
planet(<http://en.wikipedia.org/wiki/Pluto>,AD[?year]) :-
   lessThanOrEqual(1930,?year),
   lessThanOrEqual(?year,2006).
```

As part of the formal rule knowledge, the Pluto URI is employed only for naming. The rule can also be employed as metadata about informal knowledge through ('semantic search engine') queries like `planet(?which,2005)`, because one of its solutions will bind `?which` to the URI, whose dereferencing ('clicking') will then retrieve Pluto's Wikipedia entry.

Third, referring to the terminology in section E3, for certain formal purposes a URI like `http://termeg.org#MiniVan` is needed just to provide a name; for other formal purposes, also to provide a total or partial definition found by using that same URI access-style (say, the partial definition of being `rdfs:subClassOf` both `http://termeg.org#Van` and `http://termeg.org#PassengerVehicle`).

Conclusion

Four Web rule issues I1-4 led to corresponding essentials E1-4. These Web rule essentials are variously interrelated. For instance, a Rule Wiki (E1) for assertional knowledge can be extended with terminological knowledge (E3), both of which can be kept in distributed modules (E2) accessed by URIs (E4), where terminological classes may "#"-extend their module URIs for global naming (and for access to their definitions).

While the focus of this paper was on declarative rules for knowledge derivation, the four essentials can be *transferred* to (re)active rules for knowledge update, which have been increasingly studied in Web languages such as Reaction RuleML [PKB07] and Prova [Pas07]: The new event and action parts of these rules can also be beneficially combined with Controlled English (E1), modules are even more important for containing side-effects (E2), terminologies can be directly added to formalize both event and action vocabularies (E3), and all kinds of URIs are also crucial to (re)active Web rules and Web Services (E4).

Taken together, the Web rule essentials constitute a diamond-like system, $E2_4^1 3$, with URIs (E4) at the bottom, modules (E2) and assertional-terminological layers (E3) on the same level in the middle, and Controlled English (E1) at the top. The $E2_4^1 3$ system can be *augmented* by other design principles such as support for (Semantic) Web Services, APIs to (Web) databases, and interfaces with Web

2.0 tools. This paper is a kind of progress report at understanding $E2^1_43$, and further research will be needed to accelerate the evolution of Web rules as a natural extension to other kinds of Web information.

Acknowledgements

Thanks go to David Hirtle for developing the RuleML XML Schema Definitions (XSDs) and for suggesting to take advantage of wiki lists to separate each rule premise, which is now part of the Rule Wiki proposal in section E1. Further thanks go to Gregory Sherman for various Rule Wiki suggestions. Also, thanks to Benjamin Craig for extending OO jDREW and, along with Tshering Dema, for validating the examples in appendix A. Moreover, NSERC is thanked for its support through a discovery grant.

References

[ADG$^+$05] Antoniou, G., Damasio, C.V., Grosof, B., Horrocks, I., Kifer, M., Maluszynski, J., Patel-Schneider, P.F.: Combining Rules and Ontologies – A Survey. Deliverables I3-D3, REWERSE (March 2005), http://rewerse.net/deliverables/m12/i3-d3.pdf

[BAV04] Bassiliades, N., Antoniou, G., Vlahavas, I.P.: A Defeasible Logic Reasoner for the Semantic Web. In: Antoniou, G., Boley, H. (eds.) RuleML 2004. LNCS, vol. 3323, pp. 49–64. Springer, Heidelberg (2004)

[BBH$^+$05] Ball, M., Boley, H., Hirtle, D., Mei, J., Spencer, B.: The OO jDREW Reference Implementation of RuleML. In: Adi, A., Stoutenburg, S., Tabet, S. (eds.) RuleML 2005. LNCS, vol. 3791, pp. 218–223. Springer, Heidelberg (2005)

[BBY04] Bhavsar, V.C., Boley, H., Yang, L.: A Weighted-Tree Similarity Algorithm for Multi-Agent Systems in e-Business Environments. In: Proc. Business Agents and the Semantic Web (BASeWEB) Workshop. Also in: Computational Intelligence (November 2004)

[BHC05] Baisley, D.E., Hall, J., Chapin, D.: Semantic Formulations in SBVR. In: Rule Languages for Interoperability. W3C, (2005)

[BHV06] Biletskiy, Y., Hirtle, D., Vorochek, O.: Toward the Identification and Elimination of Semantic Conflicts for Integration of Ontologies. In: Canadian Semantic Web. Semantic Web and Beyond: Computing for Human Experience, pp. 135–142. Springer, Heidelberg (2006)

[BK07] Boley, H., Kifer, M.: RIF Core Design. In: W3C Working Draft, W3C (March 2007), http://www.w3.org/TR/2007/WD-rif-core-20070330/

[BKPP07] Boley, H., Kifer, M., Pătrânjan, P.-L., Polleres, A.: Rule Interchange on the Web. In: Reasoning Web 2007. LNCS, vol. 4636, pp. 269–309. Springer, Heidelberg (2007)

[BL01] Berners-Lee, T.: Webizing Existing Systems. In: World Wide Web Consortium, personal notes on: Design Issues – Architectural and Philosophical Points (May 2001), http://www.w3.org/DesignIssues/Webize.html

[BL06] Berners-Lee, T.: World Wide Web Consortium. In: personal notes on: Design Issues – Architectural and Philosophical Points (March 2006), http://www.w3.org/DesignIssues/Notation3

[BLM94]   Bugliesi, M., Lamma, E., Mello, P.: Modularity in Logic Programming. Journal of Logic Programming 19/20, 443–502 (1994)

[Bol99]   Boley, H.: ONTOFILE: Exterior and Interior Ontologies of File/HTTP URLs. In: Jaakkola, H., Kangassalo, H., Kawaguchi, E. (eds.) Information Modelling and Knowledge Bases X, IOS Press, Amsterdam, Frontiers in Artificial Intelligence and Applications, Spring (1999)

[Bol03]   Boley, H.: Object-Oriented RuleML: User-Level Roles, URI-Grounded Clauses, and Order-Sorted Terms. In: Schroeder, M., Wagner, G. (eds.) RuleML 2003. LNCS, vol. 2876, pp. 1–16. Springer, Heidelberg (2003)

[Bol06]   Boley, H.: The RuleML Family of Web Rule Languages. In: Alferes, J.J., Bailey, J., May, W., Schwertel, U. (eds.) PPSWR 2006. LNCS, vol. 4187, pp. 1–17. Springer, Heidelberg (2006)

[FHK⁺05]   Fuchs, N.E., Höfler, S., Kaljurand, K., Rinaldi, F., Schneider, G.: Attempto Controlled English: A Knowledge Representation Language Readable by Humans and Machines. In: Eisinger, N., Małuszyński, J. (eds.) Reasoning Web. LNCS, vol. 3564, pp. 213–250. Springer, Heidelberg (2005)

[GCG⁺06]   Grau, B.C., Calvanese, D., Giacomo, G.D., Horrocks, I., Lutz, C., Motik, B., Parsia, B., Patel-Schneider, P.F.: OWL 1.1 Web Ontology Language Tractable Fragments. In: W3C Member Submission (December 2006), http://www.w3.org/Submission/owl11-tractable/

[Gro04]   Grosof, B.N.: Representing e-Commerce Rules via Situated Courteous Logic Programs in RuleML. Electronic Commerce Research and Applications 3(1), 2–20 (2004)

[Hal06]   Halpin, H.: Identity, Reference, and Meaning on the Web. WWW2006 Workshop on Identity, Reference, and the Web May (2006), http://www.ibiblio.org/hhalpin/irw2006/,

[Hay04]   Hayes, P.: RDF Semantics. W3C Recommendation (February 2004)

[HF06]   Haemmerlé, R., Fages, F.: Modules for Prolog Revisited. Rapport de recherche No. 5869, INRIA, Rocquencourt March (2006), http://hal.inria.fr/inria-00070157/en/,

[Hir06]   Hirtle, D.: TRANSLATOR: A TRANSlator from LAnguage TO Rules. In: Canadian Symposium on Text Analysis (CaSTA), Fredericton, Canada, October 2006, pp. 127–139 (2006)

[HPSB⁺04]   Horrocks, I., Patel-Schneider, P.F., Boley, H., Tabet, S., Grosof, B., Dean, M.: Semantic Web Rule Language (SWRL). W3C Member Submission (May 2004), http://www.w3.org/Submission/2004/SUBM-SWRL-20040521/

[KdBBF05]   Kifer, M., de Bruijn, J., Boley, H., Fensel, D.: A Realistic Architecture for the Semantic Web. In: Adi, A., Stoutenburg, S., Tabet, S. (eds.) RuleML 2005. LNCS, vol. 3791, pp. 17–29. Springer, Heidelberg (2005)

[Knu84]   Knuth, D.E.: Literate programming. The Computer Journal 27(2), 97–111 (1984)

[Kuh07]   Kuhn, T.: AceRules: Executing Rules in Controlled Natural Language. In: Marchiori, M., Pan, J.Z., de Sainte Marie, C. (eds.) RR. LNCS, vol. 4524, pp. 299–308. Springer, Heidelberg (2007)

[LBBM06]   Li, J., Boley, H., Bhavsar, V.C., Mei, J.: Expert Finding for eCollaboration Using FOAF with RuleML Rules. In: Montreal Conference of eTechnologies 2006, pp. 53–65 (2006)

[LR98]   Levy, A.A., Rousset, M.-C.: CARIN: A Representation Language Combining Horn Rules and Description Logics. Artificial Intelligence 104(1–2), 165–209 (1998)

[MB05]   Maclachlan, A., Boley, H.: Semantic Web Rules for Business Information. In: WTAS 2005. Proc. International Conference on Web Technologies, Applications, and Services, Calgary, Canada, IASTED (2005)

[McC93]   McCarthy, J.: Notes on Formalizing Context. In: IJCAI, pp. 555–562 (1993)

[MLB07a]   Mei, J., Lin, Z., Boley, H.: ALC$_P^u$: An Integration of Description Logic and General Rules. In: Marchiori, M., Pan, J.Z., de Sainte Marie, C. (eds.) RR 2007. LNCS, vol. 4524, pp. 163–177. Springer, Heidelberg (2007)

[MLB$^+$07b] Mei, J., Lin, Z., Boley, H., Li, J., Bhavsar, V.C.: The Datalog^{DL} Combination of Deduction Rules and Description Logics. Computational Intelligence 23(3), 356–372 (2007)

[MPG07]   Muñoz, S., Pérez, J., Gutiérrez, C.: Minimal Deductive Systems for RDF. In: Franconi, E., Kifer, M., May, W. (eds.) ESWC 2007. LNCS, vol. 4519, pp. 53–67. Springer, Heidelberg (2007)

[Pas07]   Paschke, A.: Rule-Based Service Level Agreements – Knowledge Representation for Automated e-Contract, SLA and Policy Management. IDEA Verlag GmbH, Munich, forthcoming (2007)

[PKB07]   Paschke, A., Kozlenkov, A., Boley, H.: A Homogenous Reaction Rule Language for Complex Event Processing. In: EDA-PS 2007. Proc. 2nd International Workshop on Event Drive Architecture and Event Processing Systems, Vienna, Austria, September 2007 (2007)

[Rei88]   Reiter, R.: On Integrity Constraints. In: Vardi, M.Y. (ed.) Proceedings of the Second Conference on Theoretical Aspects of Reasoning about Knowledge, pp. 97–111. Morgan Kaufmann, San Francisco (1988)

[Ros06]   Rosati, R.: The Limits and Possibilities of Combining Description Logics and Datalog. In: Eiter, T., Franconi, E., Hodgson, R., Stephens, S. (eds.) RuleML, pp. 3–4. IEEE Computer Society Press, Los Alamitos (2006)

[Sch05]   Schild, K.: On the Model Theory of RDF. In: Bab und T. Noll, S. (ed.) Models and Human Reasoning – Eine Festschrift für Bernd Mahr, pp. 189–206. Wissensch. & Technik Verlag, Berlin (2005)

[SD02]   Sintek, M., Decker, S.: TRIPLE – A Query, Inference, and Transformation Language for the Semantic Web. In: ISWC 2002. 1st International Semantic Web Conference, Sardinia, Italy (June 2002)

[SvESH07] Sintek, M., van Elst, L., Scerri, S., Handschuh, S.: Distributed Knowledge Representation on the Social Semantic Desktop: Named Graphs, Views and Roles in NRL. In: Franconi, E., Kifer, M., May, W. (eds.) ESWC 2007. LNCS, vol. 4519, pp. 594–608. Springer, Heidelberg (2007)

[YK03]   Yang, G., Kifer, M.: Reasoning about Anonymous Resources and Meta Statements on the Semantic Web. In: Spaccapietra, S., March, S., Aberer, K. (eds.) Journal on Data Semantics I. LNCS, vol. 2800, pp. 69–97. Springer, Heidelberg (2003)

A   Serializations in RuleML/XML

This appendix serializes the paper's central POSL rules in RuleML/XML 0.91. Extended versions are maintained online.[6]

---

[6] http://www.ruleml.org/usecases/essentials

XML `schemaLocation` attributes point to the XSDs of the most specific existing RuleML sublanguages that still validate the instances.

POSL's "*conclusion* :- *premises* ." in our RuleML/XML basically becomes (primes indicate recursive transforms): `<Implies>` *premises' conclusion'* `</Implies>`

A.1   The Example from E1

```
<?xml version="1.0" encoding="UTF-8"?>
<RuleML
xmlns="http://www.ruleml.org/0.91/xsd"
xmlns:xsi=
"http://www.w3.org/2001/XMLSchema-instance"
xsi:schemaLocation=
"http://www.ruleml.org/0.91/xsd
http://www.ruleml.org/0.91/xsd/datalog.xsd"
>
    <Assert mapClosure="universal">
        <Implies>
            <And>
                <Atom>
                    <Rel>shipment</Rel>
                    <slot>
                        <Ind>cargo</Ind>
                        <Var/>
                    </slot>
                    <slot>
                        <Ind>price</Ind>
                        <Var>cost1</Var>
                    </slot>
                    <slot>
                        <Ind>source</Ind>
                        <Var>A</Var>
                    </slot>
                    <slot>
                        <Ind>dest</Ind>
                        <Var>B</Var>
                    </slot>
                </Atom>
                <Atom>
                    <Rel>shipment</Rel>
                    <slot>
                        <Ind>cargo</Ind>
                        <Var/>
                    </slot>
                    <slot>
                        <Ind>price</Ind>
                        <Var>cost2</Var>
                    </slot>
                    <slot>
                        <Ind>source</Ind>
                        <Var>B</Var>
                    </slot>
                    <slot>
                        <Ind>dest</Ind>
                        <Var>A</Var>
                    </slot>
                </Atom>
                <Atom>
                    <Rel>add</Rel>
                    <Var>cost</Var>
                    <Var>cost1</Var>
                    <Var>cost2</Var>
                </Atom>
            </And>
            <Atom>
```

```
                <Rel>reciship</Rel>
                <Var>cost</Var>
                <Var>A</Var>
                <Var>B</Var>
            </Atom>
        </Implies>
    </Assert>
</RuleML>
```

A.2   Two Examples from E2

The loyalty module:

```
<?xml version="1.0" encoding="UTF-8"?>
<RuleML
xmlns="http://www.ruleml.org/0.91/xsd"
xmlns:xsi=
"http://www.w3.org/2001/XMLSchema-instance"
xsi:schemaLocation=
"http://www.ruleml.org/0.91/xsd
http://www.ruleml.org/0.91/xsd/hornlog.xsd"
>
    <Assert mapClosure="universal">
        <Implies>
            <And>
                <Atom>
                    <Rel>premium</Rel>
                    <Var>customer</Var>
                </Atom>
                <Atom>
                    <Rel>regular</Rel>
                    <Var>product</Var>
                </Atom>
            </And>
            <Atom>
                <Rel>discount</Rel>
                <Var>customer</Var>
                <Var>product</Var>
                <Expr>
                    <Fun>percent</Fun>
                    <Data>5</Data>
                </Expr>
            </Atom>
        </Implies>
        <Implies>
            <And>
                <Atom>
                    <Rel>premium</Rel>
                    <Var>customer</Var>
                </Atom>
                <Atom>
                    <Rel>luxury</Rel>
                    <Var>product</Var>
                </Atom>
            </And>
            <Atom>
                <Rel>discount</Rel>
                <Var>customer</Var>
                <Var>product</Var>
                <Expr>
                    <Fun>percent</Fun>
                    <Data>10</Data>
                </Expr>
            </Atom>
        </Implies>
    </Assert>
</RuleML>
```

The `legality` module:

```xml
<?xml version="1.0" encoding="UTF-8"?>
<RuleML
xmlns="http://www.ruleml.org/0.91/xsd"
xmlns:xsi=
"http://www.w3.org/2001/XMLSchema-instance"
xsi:schemaLocation=
"http://www.ruleml.org/0.91/xsd
http://www.ruleml.org/0.91/xsd/folog.xsd"
>
    <Assert mapClosure="universal">
        <Implies>
          <And>
            <Atom>
                <Rel>payment</Rel>
                <Var>customer</Var>
                <Var>product2</Var>
                <Var>amount</Var>
                <Var>method</Var>
                <Var>time</Var>
            </Atom>
            <Atom>
                <Rel>fraudulent</Rel>
                <Var>customer</Var>
                <Var>method</Var>
                <Var>time</Var>
            </Atom>
          </And>
          <Neg>
            <Atom>
                <Rel>discount</Rel>
                <Var>customer</Var>
                <Var>product1</Var>
                <Var>percent</Var>
            </Atom>
          </Neg>
        </Implies>
        <Implies>
          <And>
            <Atom>
                <Rel>delivery</Rel>
                <Var>customer</Var>
                <Var>product2</Var>
                <Var>amount</Var>
                <Expr>
                   <Fun>date</Fun>
                   <Var>y1</Var>
                   <Var>m1</Var>
                   <Var>d1</Var>
                </Expr>
            </Atom>
            <Atom>
                <Rel>payment</Rel>
                <Var>customer</Var>
                <Var>product2</Var>
                <Var>amount</Var>
                <Var>method</Var>
                <Expr>
                   <Fun>date</Fun>
                   <Var>y2</Var>
                   <Var>m2</Var>
                   <Var>d2</Var>
                </Expr>
            </Atom>
            <Atom>
                <Rel>datediff</Rel>
                <Expr>
```

```
                    <Fun>days</Fun>
                    <Var>delta</Var>
                </Expr>
                <Expr>
                    <Fun>date</Fun>
                    <Var>y2</Var>
                    <Var>m2</Var>
                    <Var>d2</Var>
                </Expr>
                <Expr>
                    <Fun>date</Fun>
                    <Var>y1</Var>
                    <Var>m1</Var>
                    <Var>d1</Var>
                </Expr>
            </Atom>
            <Atom>
                <Rel>greaterThan</Rel>
                <Var>delta</Var>
                <Data>45</Data>
            </Atom>
        </And>
        <Neg>
            <Atom>
                <Rel>discount</Rel>
                <Var>customer</Var>
                <Var>product1</Var>
                <Var>percent</Var>
            </Atom>
        </Neg>
    </Implies>
  </Assert>
</RuleML>
```

A.3   The Rule Example from E3

```
<?xml version="1.0" encoding="UTF-8"?>
<RuleML
xmlns="http://www.ruleml.org/0.91/xsd"
xmlns:xsi=
"http://www.w3.org/2001/XMLSchema-instance"
xsi:schemaLocation=
"http://www.ruleml.org/0.91/xsd
http://www.ruleml.org/0.91/xsd/hornlog.xsd"
>
  <Assert mapClosure="universal">
    <Implies>
        <And>
            <Atom>
                <Rel>emission</Rel>
                <Var type="Real">V</Var>
                <Expr>
                    <Fun>CO2</Fun>
                    <Var>E</Var>
                </Expr>
            </Atom>
            <Atom>
                <Rel>weight</Rel>
                <Var type="Real">V</Var>
                <Expr>
                    <Fun>kg</Fun>
                    <Var>W</Var>
                </Expr>
            </Atom>
            <Atom>
                <Rel>emiweight</Rel>
```

```
            <Expr>
               <Fun>CAD</Fun>
               <Var type="Decimal">R</Var>
            </Expr>
            <Expr>
               <Fun>CO2</Fun>
               <Var>E</Var>
            </Expr>
            <Expr>
               <Fun>kg</Fun>
               <Var>W</Var>
            </Expr>
         </Atom>
      </And>
      <Atom>
         <Rel>registration</Rel>
         <Var type="Van">V</Var>
         <Expr>
            <Fun>CAD</Fun>
            <Var type="Decimal">R</Var>
         </Expr>
      </Atom>
   </Implies>
   <Implies>
      <And>
         <Atom>
            <Rel>emission</Rel>
            <Var type="Real">V</Var>
            <Expr>
               <Fun>CO2</Fun>
               <Var>E</Var>
            </Expr>
         </Atom>
         <Atom>
            <Rel>speed</Rel>
            <Var type="Real">V</Var>
            <Expr>
               <Fun>kmh</Fun>
               <Var>S</Var>
            </Expr>
         </Atom>
         <Atom>
            <Rel>emispeed</Rel>
            <Expr>
               <Fun>CAD</Fun>
               <Var type="Real">R</Var>
            </Expr>
            <Expr>
               <Fun>CO2</Fun>
               <Var>E</Var>
            </Expr>
            <Expr>
               <Fun>kmh</Fun>
               <Var>S</Var>
            </Expr>
         </Atom>
      </And>
      <Atom>
         <Rel>registration</Rel>
         <Var type="Car">V</Var>
         <Expr>
            <Fun>CAD</Fun>
            <Var type="Real">R</Var>
         </Expr>
      </Atom>
   </Implies>
   </Assert>
</RuleML>
```

A.4   The Example from E4

```xml
<?xml version="1.0" encoding="UTF-8"?>
<RuleML
xmlns="http://www.ruleml.org/0.91/xsd"
xmlns:xsi=
"http://www.w3.org/2001/XMLSchema-instance"
xsi:schemaLocation=
"http://www.ruleml.org/0.91/xsd
http://www.ruleml.org/0.91/xsd/hornlog.xsd"
>
 <Assert mapClosure="universal">
  <Implies>
   <And>
    <Atom>
     <Rel>lessThanOrEqual</Rel>
     <Data>1930</Data>
     <Var>year</Var>
    </Atom>
    <Atom>
     <Rel>lessThanOrEqual</Rel>
     <Var>year</Var>
     <Data>2006</Data>
    </Atom>
   </And>
   <Atom>
    <Rel>planet</Rel>
    <Ind
     uri=
     "http://en.wikipedia.org/wiki/Pluto"/>
    <Expr>
     <Fun>AD</Fun>
     <Var>year</Var>
    </Expr>
   </Atom>
  </Implies>
 </Assert>
</RuleML>
```

KISS – Knowledge-Intensive Service Support: An Approach for Agile Process Management

Daniela Feldkamp, Knut Hinkelmann, and Barbara Thönssen

University of Applied Sciences Northwestern Switzerland, School of Business,
Riggenbachstr. 16, 4600 Olten, Switzerland
{daniela.feldkamp,knut.hinkelmann,barbara.thoenssen}@fhnw.ch

Abstract. Automating business processes especially in the tertiary sector is still a challenge as they are normally knowledge intensive, little automated but compliance relevant. To meet these requirements the paper at hand introduces the KISS approach: modeling knowledge intensive services by enriching business rules semantically and linking rules to processes. We defined four types of rules, each focusing on a specific problem (resource allocation, constraints checking, intelligent branching and variable process planning and execution). For knowledge formalization we provide a 3-phase-procedure starting with a semi-formal representation (business model), followed by a formal representation (interchange model), leading to a machine executable representation (execution model).

Keywords: Agile Business Process, Process Modelling, Business Rules, Business Rules Formalization, Variable Process.

1 Introduction

Business process management has been very successful for structured processes with the objectives of process optimization, quality management, implementation of business information systems, or workflow management. In actual applications, however, we still face various problems: Often process documentations are not in line with the real work in the organization, e.g. because the processes are not implemented as documented or because processes have changed and the documentation is not adjusted. Also, process definition often lack the right level of granularity, i.e. they are very detailed forcing participants to follow a rigid regime and prohibiting flexibility in process execution.

The situation is even worse for knowledge-intensive or dynamic processes as they have to deal with

- exceptional situations
- unforeseeable events
- unpredictable situations, high variability
- highly complex tasks.

As Adams, Edmond and ter Hofstede [2] state are WfMSs designed to support rigidly structured business processes but not knowledge intensive processes which are

A. Paschke and Y. Biletskiy (Eds.): RuleML 2007, LNCS 4824, pp. 25–38, 2007.

by nature weakly structured and do not match at least the one crucial condition for process automation: A high repeatability rate, i.e. doing the same thing in the same way many times. Consider for example a claim process in a company: Approving a claim for damages may vary from simply checking whether the loss is covered by the insurance contract to complex evaluation including locally inspection, verification by an expert, examination of the damaged goods, checking on beneficiaries to avoid fraud etc.

If possible at all, a process model covering all possible cases would be to complex to be manageable. Typically, decisions on those actions like who has to perform the task, in which order by when, with what result etc. requires knowledge and experience and therefore are taken by a domain expert. As the process varies very much, it is normally not automated and often not even documented in detail.

Disadvantages of that situation are lack of transparency and traceability of work, inconsistent decisions and not taking into account company's regulations. Especially the increasing demands on governance and compliance have been forcing companies as well as public administrations in the last few years to review these kinds of processes for improvement.

That brings the subject of 'Business Rules' into the picture. Business rules allow for an explicit and consistent specification of regulations [16]. They provide an excellent means of encapsulating knowledge about the intentions of business and their implementation in business processes and information systems [13].

In this paper we present a new approach for agile business process management combining the strengths of both business processes modelling and business rules. Business process models are kept as simple as possible but complex enough to allow for process optimization and automation. To cope with the requirements for flexibility, consistency and compliance we integrate process models with business rules. Making consistent business rules available for all stakeholders of a business process (that is suppliers, employees, partners, clients etc.) reduces the drawback of process and workflow management when applied to knowledge intensive tasks.

2 The KISS Approach

The approach introduced in this paper combines business rules and process models in order to automate knowledge intensive services by taking advantage of both fields:

- process models are used for
  - explicit documentation and visualisation
  - execution automation.
- rules are used for
  - resource allocation
  - constraint checking
  - branching (of processes) and decision making
  - process identification and selection

To allow for flexible process execution we introduced a new modelling construct that we call *variable activity*. A variable activity is closely related to a knowledge-intensive task as introduced by Abecker et al. [1]. A variable activity corresponds to a sub process with the particularity that the activities of the sub process are determined at run-time instead of strictly modeling them.

The following example introduces the approach: In an insurance company the process for claim checking consists of five activities:

1. check claim for formal correctness
2. check claim for consistency with contract
   and decide whether to
3. rejected claim or
4. evaluated claim in detail
5. decide on claim.

Note that activity 4 (evaluate claim) is a variable activity. Depending on the specific claim one or more activities or sub-processes have to be executed. It can range from simply checking whether the loss is covered by the insurance contract to complex evaluation including locally inspection, verification by an expert, examination of the damaged goods, checking on beneficiaries to avoid fraud etc. Thus, the process model consists of activities that are fixed (1, 2, 3 and 5) and one activity (4) that is variable and will be determined not until runtime.

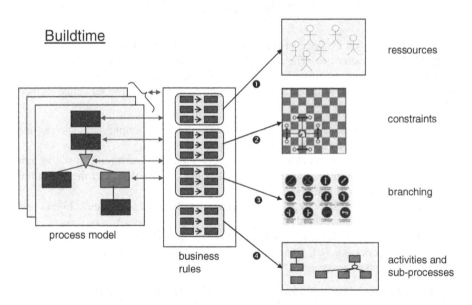

Fig. 1. KISS Build time Activities

Fig. 1 depicts the build time activities for knowledge intensive claim checking process. The process model consists of the five activities. The blue coloured are the pre-defined activities whereas the green one indicates the variable activity. The business rules are grouped for resource allocation (1), constraint checking (2), branching decisions (3) and activity or sub-process identification and selection (4). Business rules are linked to the process model. Note that same business rules can be used by several activities and processes.

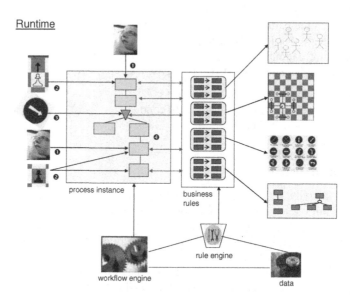

Fig. 2. KISS Runtime Activities

Fig. 2 illustrates an instance of the above process model - the variable activity (4) is expanded to two activities. The process instance is executed by a workflow engine that triggers the rule engine. Both workflow and rule engine have access to application, process and domain data. During runtime the linked rule sets are executed and

(1) the appropriate resources are allocated (the example just shows one domain expert allocated to two activities)
(2) constraints are checked (for example whether the accepted claim does not exceed the insurance sum or whether the client belongs to the platinum, gold or silver category)
(3) branching is fixed (e.g. as it is a platinum client the tolerance is higher and the claim is accepted) and
(4) the variable activity is expanded by determining the concrete activities or sub-processes (e.g. examination of the damaged goods and computing the amount of coverage).

This small example illustrates how knowledge intensive services can be modeled in a way that is flexible but is still eligible for automation.

In the following chapters we describe how to model knowledge intensive processes in the outlined way and how to execute these processes automatically.

3   KISS in Depth

Business process management is one approach to make business agile. In their seminal book on Business Reengineering, Hammer and Champy [6] mentioned adaptivity and changes as important challenges of business. This in the end led to workflow management systems. However, it was realized quite soon, that workflow management systems are mainly useful for production oriented business processes, because these processes are well-structured. But, they lack flexibility in process execution. To achieve more flexibility also for weakly structured processes, the workflow management systems were advanced toward adhoc workflows, in which the participants can modify the workflow at run time. The flexibility is achieved by an intervention of a human. There are only a few approaches toward system-supported agility [20], [10]. But these approaches are limited in the sense that they rely on predefined workflow or activities. They mainly deal with modifying the logical structure of the workflow neglecting aspects like flexible resource allocation or decision making.

An alternative approach towards agility takes into account that every business application is based on rules to conduct the business logic [19], [15], [17]. When compliance requirements increased, along with the demand for business flexibility the business rules approach emerged.

To combine the advantages of well-structured processes with the demand for adaptivity and flexibility, the integration of business rules with business process management can be regarded as promising.

Roger T. Burlton expressed: "Separate the *know* from the *flow*." (cited by Ross [15]) The implication is that the know part and the flow part are different. While business processes specify a process flow that has to be followed, e.g. for reasons of efficiency and traceability, business rules support flexibility and customization. In addition to intelligent branching and deriving values for application data (see Xpert.ivy [14] as an example) rules can be utilized to allow for context-specific adaptation of process steps. For instance, instead of assigning workflow participants because of very general roles, a rule-based approach would be able to select participants because of individual skills (e.g. select the person that had formerly solved a similar task). Additionally, separating the know from the flow has the advantage to achieve adaptivity and flexibility. In this chapter we first introduce briefly business rules and then we describe a 3-phase procedure to model agile processes combining business process modelling (the "flow") and business rules (the "know"). After it, we introduce the KISS approach for relating rules to activities.

3.1 Business Rules

The Business Rules Group [4] defines a business rule as "a statement that defines or constraints some aspect of the business. It is intended to assert business structure or to control or influence the behavior of the business."

There are several classification schemata for Business rules formalization, like Endl[5], Herbst [7], Ross [15], [16] and von Halle[19]. Since it is well understood and comprehensive for KISS we use the classification given by Barbara von Halle [19]. The rule classification depicts business rules into terms, facts and rules.

A term is a phrase which has a specific meaning for the business in some context; a fact is a statement which expresses relationships between two or more terms. Rules are a declarative statement that applies logic or computation to information values. Following the definitions of von Halle, rules are split into five sub-classes:

- **Mandatory constraints**
  Mandatory constraints are statements which have to be kept at any time. To express this kind of rules the auxiliary verb "must" is used.
- **Guidelines**
  Guidelines are rules, which could be followed but must not. The auxiliary verb "should" is used to express guidelines.
- **Action enabling**
  Action enabling rules trigger another rule or a process (step), if the condition holds.
- **Computations**
  Computations are rules to perform calculations, like sum and difference.
- **Inferences**
  Inferences are statements which establish the truth of a new fact, if the condition holds.

3.2 Agile Process Modelling Approach

To make the "know" explicit we introduce a 3-phase procedure to model adaptive processes and business rules. The first phase deals with capturing rules, drafting a process model and linking rules and processes. In the second the rules and process model are transformed into an interchange form and in the last phase the rules and the process model are transformed into a machine executable form.

3.2.1 Phase 1

The first phase of designing agile processes focuses on knowledge capturing from business users. In general, it is easier for business people to sketch a process

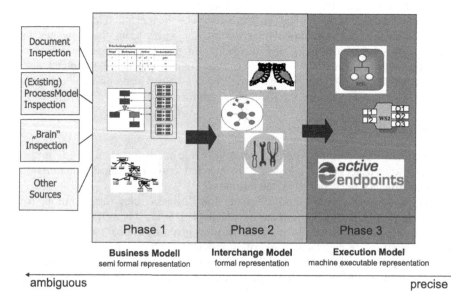

Fig. 3. Agile Process Modelling Approach

model – how business should be done – so it is recommended to start with a 'process skeleton' (q.v. Step 1). Business people normally express regulations in natural language that is by nature ambiguous, often not very precise, may be inconsistent, incomplete etc. Therefore the focal point of rule capturing is to phrase rules in a way that overcomes the drawback of natural language problems but is still well understandable.

Capturing rules and modelling the process is regarded as an iterative procedure as in the beginning it is often not very clear what should be modeled as an activity (atomic process) and which part should be better modeled as business rules.

3.2.2 Phase 2

The second phase of designing adaptive processes is focused on transformation of the models into a precise, machine understandable form.

The purpose of this interchange model derived in this phase is twofold: First, the rule of the business models are in a format that can be easily understood by business people. Therefore a semi-structured representation was chosen that contains natural language text. This, however, has the disadvantage that the rules can be ambiguous. In order to be validated and executed, the business rules and processes have to be represented in a language with well-defined semantics. Second, there can be different run-time environments for the execution of business processes. The interchange format shall serve as a common language from which the execution formats can be derived unambiguously, if possible even automatic.

To fulfill these purposes the interchange format must have a clear and precise semantics. We have pointed out, that business rules systems only have simple formalism with weak semantic for representing facts and terms. Extending the expressiveness towards ontologies has the advantage of higher expressiveness and the chance to use inferences like inheritance and consistency checking. As consequence, since the procedural knowledge must be highly integrated with declarative knowledge, a rule language must be available in which all rule types can be expressed and which can be integrated with ontologies. Therefore, we use OWL [14] and SWRL [8] to express terms, facts and business rules, while the process models are represented in OWL-S [12]. Because of the partially ambiguous business models this procedure is not automated: the process model has to be transferred into OWL-S and the rules into OWL respectively into SWRL manually. However, the semi-structured representation of the business models the development of a semantic representation is straightforward.

3.2.3 Phase 3

In the last phase the interchange model created in the second phase will be automatically migrated into machine executable forms if necessary. For example OWL-S can be transformed into BPEL and can then be executed by a workflow engine.

3.3 Relation Between Rules and Activities

In the last section we describe a 3-phase procedure for separating the "know" from the "flow". Separating business rules from activities and storing the rules independent from the processes in a repository implements the separation of know and flow (see introduction of section 0) has the advantages, that rules can be modified independent from the business logic, thus leading to more flexible and adaptable business processes. Additionally, since a single rule can be associated to several activities, business logic is more reusable.

We identified four different ways of relating rules to activities:

- Variable process execution for flexibility
- Intelligent resource allocation at run time
- Constraint checking
- Intelligent branching and decision making

This means, that an activity of a process can have relations to four different rule sets, one rule set for each of the above mentioned relation types.

As the benefits of separating but linking business process and business rules are obvious, leading software producers like ILOG and IBM already implemented an interface for example for constraint checking and decision making [9]. However, not resource allocation and least of all variable process execution have been investigated in depth. In the following the four types of relations are explained, but focussed on the variability of processes.

3.3.1 Variable Process Execution for Flexibility

Each variable activity is related to a set of action-enabling rules (maybe together with inference and computation rules), that select the appropriate activities and sub-processes. *This rule set is invoked first* to dynamically determine and instantiate the appropriate actions. They trigger planning and refinement of a process by selecting from predefined activities e.g. for requesting additional documents or notifying other administration units.

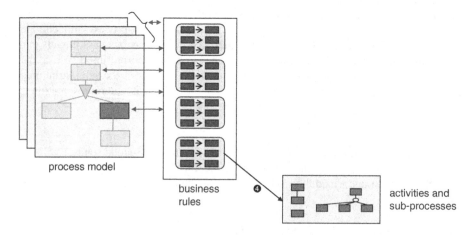

Fig. 4. Variable process execution

Regarding the claim example introduced in chapter 0, the fourth activity "evaluate claim" is a variable process, containing the following four activities:

(A) Request documents
(B) CallInsuranceSurveyor
(C) Conduct Interview
(D) FileWitnessReport

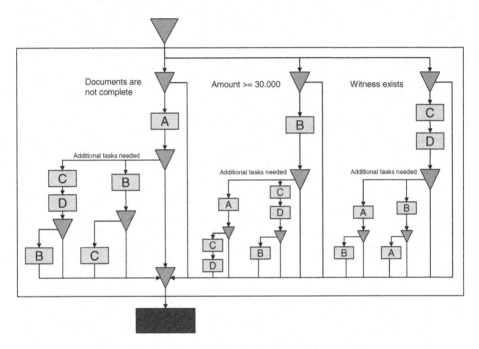

Fig. 5. Traditional process models

Fig. 5 illustrates the process in the traditional way. For example to approve a application it may be necessary to conduct several checks including requesting documents (activity A), interviewing witnesses (sub process containing a sequence of the two activities C and D) and calling insurance surveyor (activity B). All tasks depend on each other. The control flow expresses that the outcome of one task (e.g. A) may require the involvement of the other task (e.g. B and/or C, D). However, the outcome of one task also may be that the application will fail; then no further tests are required. Incorporating this behaviour in the control flow would increase the complexity enormously. In fact not all aspects of the control can be modeled in an appropriate way.

This variable process is replaced in the traditional model by a new object type, labeled with "KIT" (Knowledge-intensive task). This object is related to a pool of activities, which are linked to action-enabling rules. At run time the associated action-enabling rules select the activities that have to be executed depending on the actual context of the process instance.

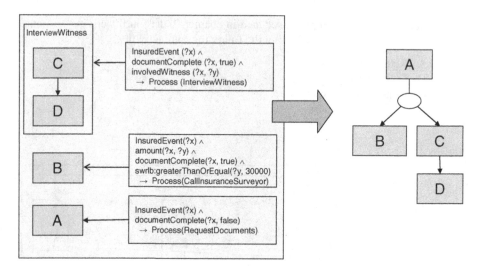

Fig. 6. Pool of sub-processes and activities

Fig. 6 shows the service model created using the KISS approach. Each of the three activities and sub processes is related to an action-enabling rule. We mentioned in section 0, that we want to use SWRL for expressing rules. To specify processes we use the OWL based Web service ontology (OWL-S). For this example we add the concepts "*InsuredEvent*" and "*Witness*". The concept "*InsuredEvent*" contains the following properties:

- *documentComplete*
  This property express if all documents are delivered, when its value is true.
- *involvedWitnesses*
  The range of the property is Witness.
- *Amount*
  The amount presents the claimed amount of the insured event.

For example, the process "*CallInsuranceSurveyor*" is executed, if the all documents are all delivered and the amount is higher than or equal 30.000.

That the outcome of an activity may require the involvement of another activity is not express explicitly. As an effect of the invocation the information can be modified and stored for example in a database. As a "side effect" further rules can be fired probably resulting in the instantiation of further activities. For example, if the client did not have sent all required documents, a request must be sent to him before executing the processes "*InterviewWitness*" and "*CallInsuranceSurveyor*". If all required documents are sent and the amount is less than 30.000, but witnesses exist, the process "*InterviewWitness*" will be invoked. Otherwise, if both conditions hold, the processes will be executed in parallel.

3.3.2 Intelligent Resource Allocation at Run Time

An activity can reference a rule set with inference and computation rules for context-specific selection of

- human resources (e.g. based on specific skills),
- information sources (e.g. using appropriate forms or adapt information presentation depending on user category) or
- IT resources (e.g. selection of a particular web service)

This rule set is applied *before invocation of the activity*.

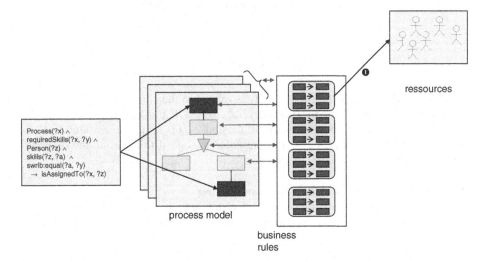

Fig. 7. Resource allocation

For the claim example, the concept *"Person"*, which contains the datatype property *"skills"* is added. The concept process of OWL-S is extended by the properties *"requiredSkills"* and *"isAssignedTo"*, which reference to *"Person"*.

The activities will be executed by persons, which skills are equal to the required skills of the process.

Even though resource allocation is a standard workflow management system functionality the KISS approach offers the opportunity of much more complex queries than for selecting appropriate roles and skills. Because of the expressiveness of business rules other sources than only the WfMS can be evaluated, e.g. yellow pages, human resource management systems and case or project databases etc. Including specific information or knowledge management systems with the help of business rules is a smooth way to increase the adequateness of allocated resources.

3.3.3 Constraint Checking

Another reference relates an activity to a rule set of mandatory constraints and guidelines (maybe together with inference and computation rules). These are checked *during run-time of the activity*.

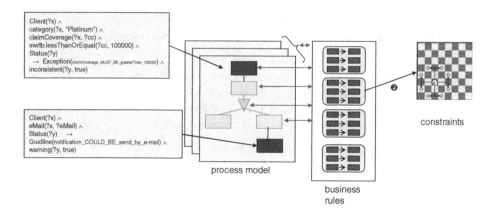

Fig. 8. Constraint checking

Regarding the example the first activity is related to the constraint, if the category of a client is "Platinum", and the claim coverage is less than or equal 100.000, then an exception must be thrown and the status of the ontology is set to inconsistent. Similar to constraints also guidelines can be represented as logical rules, using the reserved predicate "warning" instead of "inconsistent".

3.3.4 Intelligent Branching and Decision Making

Inference rules and computation rules for decision making can be associated to both activities and gateways. They derive workflow-relevant data on which the branching depends.

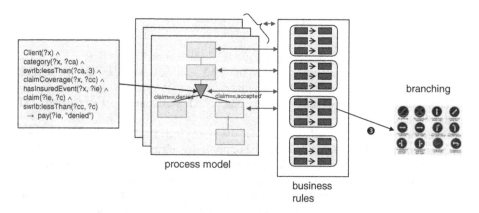

Fig. 9. Branching

Regarding the example, a gateway is related to an inference rule. If a client category is less than 3 but the claim is higher than the coverage the claim is denied. If the claim is accepted, the next activity "evaluate claim in detail" is invoked. Otherwise, if the claim is denied, the claim will be rejected.

4 Conclusion

In this paper we presented the KISS approach for agile process management. It leads to more flexible and agile business processes by integrating business process management with business rules in three different ways:

Variable activities are linked to action-enabling rules. This leads to flexible process execution, because the final process flow is determined at run-time. In contrast to ad-hoc workflows, where a human participant decides to deviate from a predefined process flow, traceability and compliance can be controlled by a set of rules.

Resource allocation is more flexible. For example, participant can be selected based on individual skills instead of general role specifications.

Intelligent branching is supported by business rules representing expert knowledge.

In addition, violation of constraints and guidelines are determined immediately at run-time avoiding expensive repair actions.

With the KISS approach, process models can be kept small and simple (following the slogan "KISS = Keep It Simple and Small") representing only the process flow while the knowledge is separated in the business rules. Separating business rules from activities and storing the rules independent from the processes in a repository has the advantages, that rules can be modified independent from the business logic. Additionally, one rule can relate to several activities, so the business logic is more reusable.

An additional advantage of our approach is achieved in that the expressiveness of business rules is extended towards ontologies allowing for inferences like inheritance and consistency checking.

The KISS approach has been developed in the FIT project[1] and is applied for e-government services of the Austrian city of Voecklabruck.

References

1. Abecker, A., Bernardi, A., Hinkelmann, K., Kühn, O., Sintek, M.: Toward a Well-Founded Technology for Organizational Memories. IEEE Intelligent Systems and their Applications 13(3), 40–48 (1998), Reprint in: Cortada, J.W., Woods, J.A(Eds.) The Knowledge Management Yearbook 1999-2000, pp. 185-199. Butterworth-Heinemann, (1998)
2. Adams, M., Edmond, D., ter Hofstede, A.H.M.: The Application of Activity Theory to Dynamic Workflow Adaptation Issues. In: PACIS-2003. The Seventh Pacific Asia Conference on Information Systems, Adelaide, Australia, pp. 1836–1852 (2003)
3. Andrews, T., Curbera, F., Dholakia, H., Goland, Y., Klein, J., Leymann, F., Liu, K., Roller, D., Smith, D., Thatte, S., Trickovic, I., Weerawarana, S.: Business Process Execution Language for Web Services (2002) (Retrieved April 4, 2007), available athttp://www-128.ibm.com/developerworks/library/specification/ws-bpel/
4. Business Rules Group: Defining Business Rules ~ What Are They Really p. 4 (2000) (Retrieved: April 19 2007), Available at: http://www.businessrulesgroup.org/first_paper/ BRG-whatisBR_3ed.pdf

[1] FIT (Fostering self-adaptive e-government service improvement using semantic Technologies) is a project funded by the European Commission within the IST program, IST-2004-27090.

5. Endl, R.: Regelbasierte Entwicklung betrieblicher Informationssysteme, Gestaltung flexibler Informationssysteme durch explizite Modellierung der Geschäftslogik. Wirtschaftsinformatik, Band 45, Josef Eul Verlag, Bern (2004)
6. Hammer, M., Champy, J.: Reengineering the Corporation. Harper Collins Publishers, Reengineering the Corporation (1993)
7. Herbst, H.: Business Rule-Oriented Conceptual Modeling. Physica-Verlag, Heidelberg (1997)
8. Horroks, I., Patel-Schneider, P., Boley, H., Tabet, S., Grosof, B., Dean, M.: SWRL - A Semantic Web Rule Language, Combining OWL and RuleML (2004) (Retrieved: June 12, 2006), http://www.w3.org/Submission/SWRL/
9. ILOG, http://www.ilog.com/
10. Karagiannis, D., Hinkelmann, K.: Context-sensitive office tasks – a generative approach. Decision Support Systems. 8, 255–267 (1992)
11. Lienhardt, H., Künzi, U.: Workflow and Business Rules: a Common Approach (2006) (Retrieved: March 18, 2007), Available at: http://www.bptrends.com/
12. Martin, D., Burstein, M., Hobbs, J., Lassila, O., McDermott, D., McIlraith, S., Narayanan, S., Paolucci, M., Parsia, B., Payne, T., Sirin, E., Srinivasan, N., Sycara, K.: OWL-S: Semantic Markup for Web Services (2004) (Retrieved: March 22, 2007), Available at: http://www.w3.org/Submission/2004/SUBM-OWL-S-20041122/
13. Morgan, T.: Business Rules and Informations Systems: Aligning IT with Business Goals. Addison-Wesley, Boston (2002)
14. McGuinness, D.L., von Harmelen, F.: OWL Web Ontology Language Overview (2004) (Retrieved: March 19, 2007), Available at: http://www.w3.org/TR/owl-features/
15. Ross, R.G.: Principles of the business rules approach. Addison-Wesley Professional, Boston (2003)
16. Ross, R.G.: Business Rules Concepts- - Getting to the Point of Knowledge. In: Ross, R.G. (ed.) Business Rule Solutions, LLC, Houston (1998)
17. Schacher, M., Grässle, P.: Agile Unternehmen durch Business Rules. Springer, Heidelberg (2006)
18. Soreco Group, Was ist Xpert.ivy? (Retrieved April 3, 2007), Available at: http://www.soreco.ch/ivy/pro/soreco/WebSite/index.jsp?navId=Products/xpertivy/was_ist_xivy
19. Von Halle, B.: Business Rules Applied, Building Better Systems Using the Business Rules Approach. Wiley and Sons, New York (2002)
20. Wargitsch, C., Wewers, T., Theisinger, F.: WorkBrain- Merging Organizational Memory and Workflow Management Systems. In: Abecker, A., Decker, S., Hinkelmann, K., Reimer, U. (eds.) Knowledge-Based Systems for Knowledge Management in Enterprises. Workshop held at the 21st Annual German Conference on AI, Document D-97-03, DFKI Kaiserslautern (1997)

Specifying Process-Aware
Access Control Rules in SBVR

Stijn Goedertier[1], Christophe Mues[2], and Jan Vanthienen[1]

[1] Department of Decision Sciences and Information Management,
Katholieke Universiteit Leuven, Belgium
{stijn.goedertier,jan.vanthienen}@econ.kuleuven.be
[2] School of Management, University of Southampton, United Kingdom
c.mues@soton.ac.uk

Abstract. Access control is an important aspect of regulatory compliance. Therefore, access control specifications must be process-aware in that they can refer to an underlying business process context, but do not specify when and how they must be enforced. Such access control specifications are often expressed in terms of general rules and exceptions, akin to defeasible logic. In this paper we demonstrate how a role-based, process-aware access control policy can be specified in the SBVR. In particular, we define an SBVR vocabulary that allows for a process-aware specification of defeasible access control rules. Because SBVR does not support defeasible rules, we show how a set of defeasible access control rules can be transformed into ordinary SBVR access control rules using decision tables as a transformation mechanism.

Keywords: access control, defeasible logic, RBAC, SBVR, BPM.

1   Introduction

Access control is the ability to permit or deny access to physical or informational resources. It is an organization's first line of defence against unlawful or unwanted acts. Recently, access control has become a key aspect of regulatory compliance. The Sarbanes-Oxley Act [1], for instance, has led to far-reaching changes in access control policies of organizations world-wide. Effective access control is managed centrally and is process aware. A centralized specification and management of access control policy is more likely to prevent flaws in the access control policy and allows to incorporate user life cycle management. In addition, access control can only safeguard the integrity of an organization's business processes when it is aware of these underlying business processes.

The design and maintenance of an enterprise-wide, process-aware access control policy is however non-trivial. Consider, for instance, a credit approval process. A customer applies for credit and after a credit review, the bank can either make a credit proposal or reject the credit application. Credit approval requires the collaboration between the sales and the risk department. Suppose, for instance, that the following access control policy is formulated:

A. Paschke and Y. Biletskiy (Eds.): RuleML 2007, LNCS 4824, pp. 39–52, 2007.
© Springer-Verlag Berlin Heidelberg 2007

stakeholders: the customer, the bank, regulators
threat: the bank accepts credit applications with a high probability of default.
threat: credit reviews take up too much time and the customer defects to another bank.
concern: In general, each credit application must be reviewed by the bank's risk department.
concern: For reasons of efficiency, credit applications of less or equal than 2000 euros may be reviewed by the sales department.
concern: The employee who reviews a credit application can neither be the beneficiary nor the applicant of the credit application.
concern: The same employee should not both review a credit application and make a credit proposal for credit applications larger than 2000 euros.

To date many access control specifications are either process agnostic or process driven. **Process-driven** access control specifications, hinder both design and run-time flexibility, because they are embedded within procedures, applications or process models. For instance, the left-hand side of Fig. 1 embeds the access control policy in a BPMN decision gateway [2]. Such process-driven specifications, hinders traceability and raises a myriad of problems when the policy is duplicated in multiple implementation forms. Duplication and lack of traceability of access control policies undermine the ability to have a consistent, flexible and guaranteed enterprise-wide access control policy. The opposite situation also entails problems. **Process-agnostic** access control specifications have only limited expressiveness, because they cannot relate to the state of business processes to grant or deny access rights. For instance, the above-defined policy prevents the same employee from both reviewing and making a proposal. This cannot be expressed without an awareness for an underlying process model. Expressive and flexible access control specifications are **process-aware** in that they can refer to an underlying business process context, but do not specify when and how they must be enforced. Such a specification is represented in the right-hand side of Fig. 1.

Process-aware access control policies should be on the one hand comprehensible so that they can be understood by business people and on the other hand

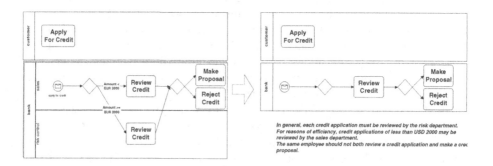

Fig. 1. Process-driven versus process-aware access control specifications

formal so that they can be enforced by information systems. The Semantics of Business Vocabulary and Business Rules (SBVR) [3,4] is a language for business modeling that has such property. In this paper we give a brief introduction to the SBVR and demonstrate how a role-based, process-aware access control policy can be specified. To this end we define a new SBVR vocabulary for verbalizing process-aware, access control rules. The vocabulary is part of a general initiative to declaratively model the business concerns that govern business processes in terms of business rules [5]. Access control specifications adhering to the role-based access control (RBAC) model are expressed in terms of general rules and exceptions and are related to defeasible logic. Because SBVR does not support defeasible rules, we show how a set of defeasible access control rules can be transformed into ordinary SBVR access control rules using decision tables as a transformation mechanism.

2   An Introduction to SBVR

The Semantics of Business Vocabulary and Business Rules (SBVR) is a new standard for business modeling that currently is under finalization within the Object Management Group (OMG). The standard provides a number of conceptual vocabularies for modeling a business domain in the form of a vocabulary and a set of rules [3,4]. In SBVR, meaning is kept separate from expression. As a consequence, the same meaning can be expressed in different ways. In real-life, meaning is more often expressed in textual form than in diagrams as statements provide more flexibility in defining vocabulary and expressing rules. For these reasons, the SBVR specification defines a structured, English vocabulary for describing vocabularies and verbalizing rules, called SBVR Structured English [4]. One of the techniques used by SBVR structured English are font styles to designate statements with formal meaning. In particular,

- the term font (normally green) is used to designate a noun concept.
- the name font (normally green) designates an individual concept.
- the *verb* font (normally blue) is used for designation for a verb concept.
- the keyword font (normally red) is used for linguistic particles that are used to construct statements.

The definitions and examples in the remainder of the text use these SBVR Structured English font styles.

In SBVR a vocabulary and a set of rules make up a so called 'conceptual schema'. A conceptual schema with an additional set of facts that adheres to the schema is called a 'conceptual model'. Figure 2 depicts the relationship of a conceptual schema and a conceptual model to some of the core building blocks in SBVR. These core building blocks are part of the SBVR 'Meaning and Representation Vocabulary'. This vocabulary contains among others the following definitions [4]:

> A conceptual schema *is* a combination of concepts and facts (with semantic formulations that define them) of what is possible, necessary, permissible, and obligatory in each possible world.

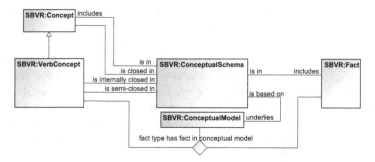

Fig. 2. A MOF/UML representation of SBVR conceptual schema and model [4]

A conceptual model or fact model *is* a combination of a conceptual schema and, for one possible world, a set of facts (defined by semantic formulations using only the concepts of the conceptual schema).

The facts in a conceptual model may cover any period of time. Changing the facts in a conceptual model creates a new and different conceptual model. In this way the SBVR gives conceptual models a monotonic semantics.

The SBVR provides a vocabulary called the 'Logical Formulation of Semantics Vocabulary' to describe the structure and the meaning of vocabulary and business rules in terms of formalized statements about the meaning. Such formalized statements are called 'semantic formulations' [6]. In addition to these fundamental vocabularies, the SBVR provides a discussion of its semantics in terms of existing, well-established formal logics such as First-Order logic, Deontic Logic and Higher-Order logic. This combination of linguistics and formal logic provides the fundamentals for developing a natural language parser that allows to express the meaning of rules that have a textual notation [7,8].

3   An SBVR Vocabulary for Process Modeling

The SBVR is a suitable base languages for defining process-aware access control rules, but to date there exists no SBVR vocabulary with process related concepts such as agents, activities and events. Consequently, it is not possible to declaratively refer to the state of a business process. In [5] we define an SBVR vocabulary for expressing process-related concepts, called the EM-BrA^{2}CE Vocabulary. EM-BrA^{2}CE stands for 'Enterprise Modeling using Business Rules, Agents, Activities, Concepts and Events'. The vocabulary thinks of a business process instance as a *trajectory* in a *state space* that consists of the possible sub-activities, events and business concepts. Each activity in a process instance can undergo a number of distinct state transitions. The occurrence of a state transition is logged as an activity event. Business rules determine whether or not a particular state transition can occur. Consider, for instance the following state transitions:

- *create*(*AId, AT, BId, PId, CoordinatorId*): requests the creation of a new activity *AId* of type *AT* with business identifiers *BId*, parent activity *PId* by an agent *CoordinatorId*. Activity event type: *created*.
- *assign*(*AId, AgentId, CoordinatorId*): requests the assignment or revocation of the assignment of activity *AId* to an agent *AgentId* by an agent *CoordinatorId*. Activity event type: *assigned*.
- *updateFact*(*AId, C_1, C_2, WorkerId*): requests the update of a business fact C_1 by C_2 within the context of activity *AId* by an agent *WorkerId*. Activity event type: *factUpdated*.
- *complete*(*AId, WorkerId*): requests the completion of activity *AId* by an agent *WorkerId*. Activity event type: *completed*.

In [5] a total of twelve generic state transitions have been identified and a generic execution model has been defined in terms of Colored Petri Nets. Figure 3 illustrates a number of state transitions that occur to a given 'review credit' activity 'a1'. Notice that each state transition results in a new set of concepts and ground facts, and thus a new state, that are partially represented in the columns of the figure. As each new activity state is considered to be a new SBVR:conceptual model, deductive reasoning can use a monotonic reasoning paradigm [4]. The current state of an activity determines which state transitions can occur. For the purpose of access control, the *assign*(*AId, AgentId, CoordinatorId*) is the activity state transition of interest.

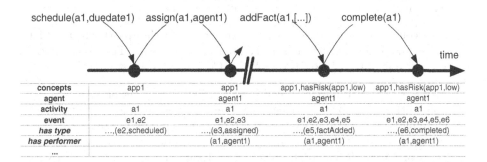

Fig. 3. An illustration of the state transitions for a 'review credit' activity 'a1'

To make this text self-contained a number of concepts of the EM-BrA^{2}CE Vocabulary are discussed in the remainder of this section. These concepts are depicted in Fig. 4. 'Activity' is a central concept in the vocabulary. An activity can either represent the act of performing an atomic unit of work or the act of coordinating a set of sub-activities (a business process). The former is called an atomic activity whereas the latter is called a composite activity. Each activity has an activity type.

An activity type *is* an SBVR:concept type that *specializes* the individual concept 'activity' and that *classifies* an activity.

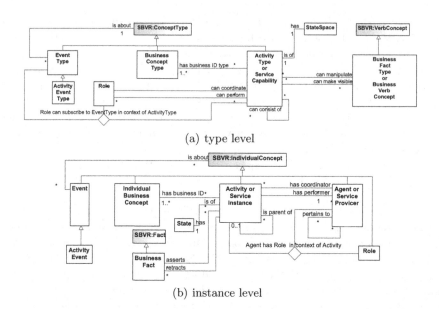

(a) type level

(b) instance level

Fig. 4. A MOF/UML representation of the EM-BrA²CE Vocabulary

An activity *is* an SBVR:individual concept that represents a unit of (coordination) work to be performed by an agent.

In the context of a business process an agent can fulfill a particular role that represents an authorization to perform a number of activities. This conception of role is consistent with the Role Based Access Control (RBAC) standard [9,10,11].

> A role *is* anSBVR:individual concept that represents a set of authorizations with regard to the performance of activities of given activity types. Example: the roles 'applicant', 'sales representative', 'risk assessor',.
> An agent *is* an SBVR:individual concept that represents an actor or a group of actors who can perform activities. Example: the agents 'workerX', 'sales department', 'banking inc.'.
> 'Agent *can have role* role' *is* an SBVR:associative fact type that represents that an agent can assume a particular role. Example: 'workerX *can have role* risk assessor'
> 'Agent *has role* role *in the context of* activity' *is* an SBVR:associative fact type that represents that an agent assumes a particular role in the context of an activity.

In the vocabulary, the state of an activity includes the history of events related to the activity or its sub-activities. Unlike many ontologies for business modeling, such as for instance the UFO [12], a distinction is made between activities and events. Activities are performed by agents and have a particular duration whereas events occur instantaneously and represent a state change in the world.

Changes to the life cycle of an activity are reflected by means of activity events. Activity events allow process modelers to distinguish between the activity state transitions that occur when, among others, creating, scheduling, assigning, starting and completing an activity.

> An event *is* an SBVR:individual concept that corresponds to an instantaneous, discrete state change of a concept in the world.
>
> An event type *is* an SBVR:concept type that *specializes* the individual concept 'event' and that *classifies* an event.

Agents that have a particular role in the context of a business process have the authorization to perform a particular activity. This authorization is expressed by the 'can perform' fact type. When performing an activity of a particular activity type, an agent can manipulate business facts of particular business fact types. This is expressed by the 'can manipulate' fact type. Additionally, agents can retrieve information about particular business fact types when performing activities. The business fact types that are visible are indicated by the 'can make visible' fact type.

> 'Role *can perform* activity type' represents the authorization that an agent that has a given role can perform an activity of a particular activity type. Note: These authorizations can be restricted by an activity authorization constraint. Example: sales representative *can perform* review credit.
>
> 'Activity type *can manipulate* business fact type' *is* an SBVR:associative fact type that represents that a business fact of type business fact type can be asserted or retracted during the performance of an activity of type activity type. Example: review credit *can manipulate* the business fact type 'credit application *has* risk'.
>
> 'Activity type *can make visible* business fact type' *is* an SBVR:associative fact type that represents the business fact types that can be made visible in the context of activities of activity type. Note: visibility can be restricted by a visibility constraint. Example: review credit *can make visible* the business fact type 'applicant *has* income'. Example: apply for credit *can make visible* the business fact type 'credit application *has* reason of rejection'

The fact type 'role *can subscribe to* event type *in context of* activity type' expresses the visibility of events to agents in the context of an activity. Notice that it does not express how agents are notified of the event, which can generally occur using either a pull, a push or a publish-subscribe mechanism. Furthermore, it is possible that the visibility is constrained by so-called 'event subscription constraints'.

> 'role *can subscribe to* event type *in context of* activity type' *is* an SBVR:associative fact type that expresses that an agent with a particular role can subscribe to an event of event type in the context of an activity of activity type. Example: An applicant *can subscribe to* completed *in the context of* review credit.

'agent *perceives* event' *is* an SBVR:associative fact type that expresses that an event is non-repudiable to a particular agent. Example: anAgentX *perceives* anEventY.

4   Specifying Access Constraints in SBVR

The EM-BrA^{2}CE Vocabulary allows to specify access control policies that are able to refer to the state of a business process instance. In particular, each business process can be modeled by describing its state space and the set of business rules that constrain movements in this state space. For instance, the state space of the credit approval process is described by facts about the following concepts:

- **roles:** applicant, sales representative, risk assessor
- **atomic activity types:** apply for credit, review credit, make proposal, reject credit
- **activity event types:** created, assigned, started, completed
- **business concepts:** credit application, amount, collateral, applicant, income, risk
- **business fact types:** credit application *has* collateral, credit application *has* amount, applicant *has* income

Access control rules constrain the $assign(AId, AgentId, CoordinatorId)$ state transition. In the EM-BrA^{2}CE Vocabulary [5] three kinds of kinds of access rules can be specified: activity authorization constraints, visibility constraints and event subscription constraints. An activity authorization constraint allows to constrain the agent-role assignments that can be granted to an agent. For instance, the fact 'sales representative *can perform* review credit' is constrained by the rule that credit applications larger than 2000 euros cannot be reviewed by employees of the sales department.

> An activity authorization constraint is a structural business rule that dynamically constrains the activities that can be assigned to an agent based on the properties of the activity, the business facts in its state space and the properties of the agent.
> Example: It is impossible that an agent that *is from department* sales department can perform a review credit activity that *has subject* credit application that *has* an amount larger than 2000 euros.
> Example: It is impossible that an agent that *has been assigned to* a review credit activity that *has subject* credit application that *has* an amount larger than 2000 euros can perform a make proposal activity that *has subject* credit application

Access control does not only refer to the performance of activities, but also deals with the access to informational resources. In the context of an activity an agent may be able to perceive particular business facts. For example, the fact 'apply for credit *can make visible* 'credit application *has* reason of rejection'' implies that agents that apply for credit can see the reason of rejection if any.

A visibility constraint is a structural business rule that dynamically constrains the visibility of business facts to agents based on the properties of the business facts and the behavior of the agents.

Example: It is not possible that the fact type 'credit application *has* reason of rejection' is visible to the agent who *has role* applicant if the credit application *has type* consumer credit.

In addition to limiting access to business facts, access rules must also be able to limit the visibility of the behavior of agents to other agents. For example, the fact 'a risk assessor *can subscribe to* started *in the context of* apply for credit'.

An event subscription constraint is a structural business rule that constrains the conditions under which agents who have a particular role in the context of an activity can perceive the occurrence of an activity event.

Example: It is not possible that an agent that *has role* risk assessor *perceives* a started event that *is about* an apply for credit activity that *has subject* a credit application that *has* an amount of less than 2000 euros.

When an event occurs, each agent who has a particular role in the context of the activity and whose role is subscribed to the event type and for whom no subscription constraints apply, can perceive the event. Consequently, the event is non-repudiable to external agents such that any legal obligation that results from the event can be enforced.

The proposed vocabulary and business rules allow to specify process-aware access control in four steps:

1. Define an **access control policy.** An access control policy is a non-actionable directive that identifies security threads and safety concerns and motivates access control implementation. Metamodels such as the OMG's Business Motivation Model [13] provide the required structure to formulate access control policies.
2. Identify the access control **roles.** Roles are permissions involving the performance of activities or the involvement in activities that pertain to meaningful groups of activity types. Roles provide *stability* because role definitions do not change as often as agent-activity type assignments would. Fact types: 'role *can perform* activity type', 'activity type *can manipulate* business fact type', 'activity type *can make visible* business fact type' and 'role *can subscribe to* event type *in context of* activity type'.
3. Make **agent-role assignments.** Agent-role assignment is the provisioning of agents with roles that represent access rights. Agent-role assignments can be defined by derivation rules, but for reasons of compliance and flexibility, agent-role assignments are often hard-coded. Fact type: 'agent *can have role* role'.
4. **Specify access constraints.** Access constraints refine the role-based access control policy to take into account issues that are beyond the scope of user-role assignment. Access constraints give an access control model *precision,*

because they constrain the role-based access according to the properties of the agent, the activity and the business process event history.

5   Verbalizing Defeasible Access Control Rules

Access control specifications adhering to the role-based access control (RBAC) have a non-monotonic semantics that can be expressed in defeasible logic [14,15]. Defeasible logic is a means to formulate knowledge in terms of general rules and exceptions. To this end, defeasible logic allows for rules of which the conclusions can be defeated (defeasible rules) by contrary evidence provided by strict rules, other defeasible rules and defeaters. A superiority relation between rules indicates which potentially conflicting conclusions can be overridden. In EM-BrA^{2}CE the 'agent can perform activity' fact type is defined using a rule set of two generic defeasible rules:

$$r_1 : true \Rightarrow \neg canPerform(G, A),$$
$$r_2 : hasRole(G, R), hasType(A, At), canPerform(R, At)$$
$$\quad \Rightarrow canPerform(G, A)$$

and a number of domain-specific activity authorization constraints, that translated to the following defeasible rules:

$$r_3 : A(3) \Rightarrow \neg canPerform(G, A),$$
$$...$$
$$r_i : A(i) \Rightarrow \neg canPerform(G, A),$$
$$...$$
$$r_n : A(n) \Rightarrow \neg canPerform(G, A).$$

The following priority relationship applies between the generic and domain-specific defeasible rules:

$$r_1 < r_2, r_2 < r_3, r_2 < r_4, ... r_2 < r_n.$$

Because activity authorization constraints all potentially defeat the 'agent can perform activity' conclusion, they cannot contradict each other. As a consequence, no activity authorization constraint can be specified that can invalidate the outcome of another constraint. This is a valuable result, because it facilitates the incremental specification of access control policies [16]: new rules can be added without the conditions of previous rules need to be reconsidered. Ordinary rules, in contrast, require a complete, encyclopedic knowledge of all rules to be updated or decided upon. In addition to their enhanced representational capability, efficient implementations of defeasible logic have been proposed [17] together with visualization techniques to render the added complexity of default reasoning comprehensible to end users [18,19].

Although defeasible logic is naturally present in the role-based access control model, the current SBVR specification does not allow for the specification of defeasible rules. To cope with this shortcoming, defeasible access control rules must be transformed into a set of ordinary, strict rules. Decision tables can be such a transformation mechanism. Decision tables are one of many visualizations [20] of rule sets. Figure 5 displays an example of a decision table. Graphically, a decision table consists of four quadrants. The two upper quadrants make up the condition sphere and the two lower quadrants represent the conclusion sphere. Properly built decision tables contain columns that represent exhaustive and mutually exclusive conjunctions of conditions and that have one or more conclusions. A set of input rules, possibly expressed using a defeasible logic, determines which combination of conditions in the upper right quadrant leads to which conclusion values in the lower right quadrant.

Several tools provide support in constructing decision tables [21,22,23,24]. In Prologa [22] decision tables are constructed by providing conditions with mutually exclusive condition labels, conclusions and input rules. Although Prologa fits in a propositional logic framework, Prologa proves a useful tool to visualize and transform a number of input rules, expressed using Prologa's own default logic, into table rules. In particular, Prologa provides the user with a number of features that facilitate knowledge modeling, such as the reordering of condition labels to expand or contract the table, the syntactical verification of rules and the helpful visualizations that allow to semantically validate a rule set. The columns of the decision table in Fig. 5 represent the six 'review credit' access rules that are obtained by transforming the three defeasible rules of the access control policy. For instance, column 6 translate in the following SBVR rule:

It is necessary that an **agent** that *is* not *applicant of* the credit application and that *is* not *beneficiary of* the credit application and that *has* not *role* risk assessor and that *has role* sales representative, can not perform

review credit						
1. agent *is applicant of* credit application	Y		N			
2. agent *is beneficiary of* credit application	-	Y	N			
3. agent *has role* risk assessor	-	-	Y	N		
4. agent *has role* sales representative	-	-	-	Y	N	
5. credit application *has* amount larger than 2000 euros	-	-	-	Y	N	-
1. agent can perform the review credit activity	-	-	x	x	-	-
2. agent cannot perform the review credit activity	x	x	-	-	x	x
	1	2	3	4	5	6

R_1 In general, conclusion 2.
R_2 In general, conclusion 1 if condition 3.
R_3 In general, conclusion 1 if condition 4.
R_4 In general, conclusion 2 if condition 1.
R_5 In general, conclusion 2 if condition 2.
R_6 In general, conclusion 2 if condition 4 and condition 5.

$R_1 < R_2 < R_3 < R_4 < R_5 < R_6$

Fig. 5. Transforming defeasible access rules into ordinary access rules

a review credit activity that *has subject* credit application that *has* an amount more than 2000 euros.

Six ordinary SBVR rules is the minimal number of rules required to capture the access control policy. Because SBVR rules do not allow defeasibility, the conditions of every activity authorization constraint must be repeated in almost every rule resulting in a larger number of longer rules.

6 Related Work

The Semantics of Business Vocabulary and Business Rules (SBVR) provides a number of conceptual vocabularies for modeling a business domain in the form of a vocabulary and a set of rules. The current SBVR specification [4] does not have a built-in vocabulary for expressing process-related concepts such as agent, activity, event or deontic assignment. Such vocabularies and formal semantics for expressing dynamic constraints are deferred to a later version of the SBVR standard [4]. Without such a vocabulary it is difficult to express process-aware access control rules. In [5] we have supplemented the SBVR with a vocabulary for declarative process modeling called the EM-BrA^{2}CE Vocabulary.

The access control model of the EM-BrA^{2}CE Vocabulary is based on the existing standard for Role-Based Access Control [9,10,11]. The existing RBAC standard allows to specify dynamic separation of duty (SoD) constraints. Such constraints impose that within a single session a user cannot assume two roles on which a dynamic SoD constraint applies. Strembeck et al. discuss an extension to the RBAC standard by allowing to express dynamic, process-related, access rules [25]. The proposed EM-BrA^{2}CE Vocabulary goes further in that it allows to express process-aware access control policies that can declaratively relate to the current state of a business process and to a history of related business events.

The SBVR does by itself not allow for the specification of default rules or defeasible rules. Nonetheless, it is very common in natural language to use a kind of default logic to express access control policies. In this paper we have applied the long-lived practice of using decision tables as a transformation mechanism to transform default rules into a set of strict rules. On the output a (minimal set) of strict rules is produced that is equivalent to the defeasible input rules. A decision table visualisation serves a different purpose than for instance an immediate visualization of defeasible rules. Bassiliades et al., for instance, describe a visualization technique for defeasible logic rules based on directed graphs of so-called literal boxes and predicate boxes. Predicate boxes are connected with different connection types indicating the kind of rule or the priority relationship [18,19]. In correspondence to the Prologa tool, this technique is implemented at the level of the literals rather than at the level of terms, including variables and variable conditions. This graph-based technique seems an intuitive, user-friendly representation of defeasible rules. Decision tables, in contrast, serve the purpose of verifying a defeasible rule set and allow to transform the rule set into ordinary rules.

7   Conclusion

In this paper we have indicated how process-aware, role-based access control specifications can benefit form the upcoming SBVR standard. In particular, an SBVR vocabulary for process modeling was defined that allows to declaratively refer to the state of a business process when specifying access control rules. This vocabulary allows to model a business process by describing its state space and the set of business rules that constrain movements in this state space. The advantage is that access control rules are process aware, yet maintain a declarative nature by not specifying how and when an access rule must be enforced. To cope with the absence of defeasible logic in SBVR, we have shown how a set of defeasible access control rules can be transformed into SBVR access control rules using decision tables as a transformation mechanism. The SBVR offers many research opportunities to support the standard both in terms of linguistics and logics.

References

1. Securities and Exchange Commission, U.S.A.: Sarbanes Oxley Act 2002. Securities and Exchange Commission (SEC), U.S.A (2002)
2. Object Management Group: Business Process Modeling Notation (BPMN) – final adopted specification. OMG Document – dtc/06-02-01 (2006)
3. Chapin, D.: Semantics of Business Vocabulary & Business Rules (SBVR) [26]
4. Object Management Group: Semantics of Business Vocabulary and Business Rules (SBVR) – Interim Specification. OMG Document – dtc/06-03-02 (2006)
5. Goedertier, S., Vanthienen, J.: EM-BrA<Superscript>2</Superscript>CE v0.2: A Vocabulary and Execution Model for Declarative Process Models. Fetew research report, K.U.Leuven (2007), http://www.econ.kuleuven.ac.be/public/ndbaf38/EM-BrAACE
6. Baisley, D.E., Hall, J., Chapin, D.: Semantic Formulations in SBVR [26]
7. Unisys: Unisys rules modeler (2005) (10-11-2005), www.unisys.com
8. Digital Business Ecosystem (DBE): Sbeaver (2007), http://sbeaver.sourceforge.net
9. Sandhu, R.S., Coyne, E.J., Feinstein, H.L., Youman, C.E.: Role-based access control models. IEEE Computer 29(2), 38–47 (1996)
10. Ferraiolo, D.F., Sandhu, R.S., Gavrila, S.I., Kuhn, D.R., Chandramouli, R.: Proposed nist standard for role-based access control. ACM Trans. Inf. Syst. Secur. 4(3), 224–274 (2001)
11. InterNational Committee for Information Technology Standards (INCITS): Role-Based Access Control. American National Standard ANSI/INCITS 359-2004 (2004), http://csrc.nist.gov/rbac
12. Guizzardi, G., Wagner, G.: Ontologies and Business Systems Analysis. In: Rosemann, M., Green, P. (eds.) Some Applications of a Unified Foundational Ontology in Business Modeling, pp. 345–367. IDEA Publisher, USA (2005)
13. Object Management Group: Business Motivation Model (BMM) – adopted specification. OMG Document – dtc/2006-08-03 (2006)
14. Nute, D.: Defeasible Logic. In: Handbook of Logic in Artificial Intelligence and Logic Programming, pp. 353–395. Oxford University Press, New York (1994)

15. Antoniou, G., Billington, D., Governatori, G., Maher, M.J.: Representation results for defeasible logic. ACM Trans. Comput. Log. 2(2), 255–287 (2001)
16. Grosof, B.N., Labrou, Y., Chan, H.Y.: A declarative approach to business rules in contracts: courteous logic programs in XML. In: ACM Conference on Electronic Commerce, pp. 68–77. ACM Press, New York (1999)
17. Maher, M.J., Rock, A., Antoniou, G., Billington, D., Miller, T.: Efficient defeasible reasoning systems. International Journal on Artificial Intelligence Tools 10(4), 483–501 (2001)
18. Bassiliades, N., Kontopoulos, E., Antoniou, G.: A visual environment for developing defeasible rule bases for the semantic web. In: Adi, A., Stoutenburg, S., Tabet, S. (eds.) RuleML 2005. LNCS, vol. 3791, pp. 172–186. Springer, Heidelberg (2005)
19. Kontopoulos, E., Bassiliades, N., Antoniou, G.: Visualizing defeasible logic rules for the semantic web. In: Mizoguchi, R., Shi, Z., Giunchiglia, F. (eds.) ASWC 2006. LNCS, vol. 4185, pp. 278–292. Springer, Heidelberg (2006)
20. Antoniou, G., Taveter, K., Berndtsson, M., Wagner, G., Spreeuwenberg, S.: A First-Version Visual Rule Language. Report IST-2004-506779, REWERSE (2004)
21. Vanthienen, J., Robben, F.: Developing legal knowledge based systems using decision tables. In: ICAIL, pp. 282–291 (1993)
22. Vanthienen, J., Mues, C., Aerts, A.: An Illustration of Verification and Validation in the Modelling Phase of KBS Development. Data Knowl. Eng. 27(3), 337–352 (1998)
23. Spreeuwenberg, S., Gerrits, R., Boekenoogen, M.: Valens: A knowledge based tool to validate and verify an aion knowledge base (2000)
24. Vanthienen, J., Mues, C.: Prologa 5.3 - tabular knowledge modeling (2005)
25. Strembeck, M., Neumann, G.: An integrated approach to engineer and enforce context constraints in rbac environments. ACM Trans. Inf. Syst. Secur. 7(3), 392–427 (2004)
26. W3C Workshop on Rule Languages for Interoperability, 27-28 April 2005, Washington, DC, USA. In: Rule Languages for Interoperability, W3C (2005)

A Rule-Based Approach to Prioritization of IT Work Requests Maximizing Net Benefit to the Business

Maher Rahmouni[1], Claudio Bartolini[2], and Abdel Boulmakoul[1]

[1] Hewlett-Packard Laboratories, Bristol, UK
[2] Hewlett-Packard Laboratories, Palo Alto, USA
name.surname@hp.com

Abstract. With the growth of IT outsourcing opportunities, providers of managed IT services, such as HP, are compelled to exploit any possible economy of scale in dealing with emerging requirements of their customers. In this paper we present a methodology and a set of tools for enhanced executive decision support capability for the best allocation of scarce development resources through timelier and more accurate delivery of forecast indicators relative to the net benefit. Examples of such indicators are total forecast revenue for a bundle of work requests and total forecast cost reductions for a bundle of work requests. The tools deliver on reduced development cost and shorter time to value through the identification of synergies, duplication and commonality in work requests. For example, our approach will be able to identify in a reliable manner work that falls into these discrete categories pertaining to potential duplication thereby highlighting areas of potential cost reduction. Moreover they guarantee a reduced turn around time of delivery to trade customers through prioritization driven by net benefit and optimized release.

Keywords: IT oursourcing, IT service management, Work request prioritization.

1 Introduction

With the growth of IT outsourcing opportunities, providers of managed IT services, such as HP, IBM, EDS and others, are compelled to exploit any possible economy of scale in dealing with emerging requirements of their customers. These requirements are expressed through *work requests*. Work requests may span both the pursuit phase (leading up to bidding for a contract) and the delivery phase (execution of terms and conditions of the outsourcing contract) of business engagement with outsourcing customers. In delivery, work requests express requests for addition of features to existing products, or for architectural review of a product or solution. In the pursue phase, they may be relative to assisting the implementation manager to develop an implementation plan[1] or a pre-Sales bid work relating to automation tools architectures.

Given that the development resources (people, technology and processes) available for executing on the work requests are finite, a big problem that the outsourcing

---

[1] Any activity relating to executing the implementation plan is out of scope.

A. Paschke and Y. Biletskiy (Eds.): RuleML 2007, LNCS 4824, pp. 53–62, 2007.

organization faces is the prioritization of work requests, so as to best allocate the development resources.

The main obstacle to an efficient and cost-effective allocation of the development resources is the difficulty in assessing the net benefit for a work request. Estimating the net benefit requires: both a precise assessment of the business impact of the work requests being prioritized and a detailed cost estimate for the development. Information substantiating cost and benefits is collected throughout the lifecycle of the work request (figure 1). Examples of substantiating information are a detailed *business requirements document* on the submitter side, and *functional specification* and a *detailed estimate* on the supplier side. However, there is tension between the need of having accurate estimate early enough in the lifecycle, and the fact that the supporting document may come only late in the lifecycle. For instance, developer teams only provide a full functional specification once a decision to go ahead with the fulfillment of the request has been made. There is also very limited or no visibility of the development teams resources (their current workload, their skill set).

If this were not a hard enough problem, complications come into the picture when one wants to take into account the dependencies among work requests. Because of complex and inexplicit code or product dependencies – developer teams tend to underestimate the effort needed to fulfill the request. From our conversation with outsourcing managers, we have heard anecdotes of when nearing a work requests deadline it suddenly dawned on the service enablers that dependent software components have to undergo change in order not to break the customers' solution. Once again, this increases the cost of delivery due to obtaining vital information late rather than early in the work request lifecycle.

Moreover, dependencies can exist between work requests that have been submitted. The prioritization process needs to consider bundles of work requests rather than work requests in isolation. For instance, when similarity is identified between two or more work requests, any estimate made on the development effort might have to be reduced, since same or complementary products and code bases will have to be modified. On the other hand, when considering bundles of work requests linked together by a common deadline, the estimate on the delivery time will have to take into account the additional workload imposed onto the development teams with respect to what would be the case if considering each work request in isolation. Similar considerations could be made that apply to the business value estimate.

This paper presents a rule-based approach to constructing and prioritizing *bundles* of dependent work requests. The remainder of this document is organized as follows: Section 2 provides an overview of the decision support solution for managing the work request lifecycle. Section 3 describes the rule based system used for modelling dependencies between work requests and bundling them together. In section 4, the prioritization scheme is presented. Finally, a discussion of techniques used in this paper is performed and further improvements to the actual solution are suggested.

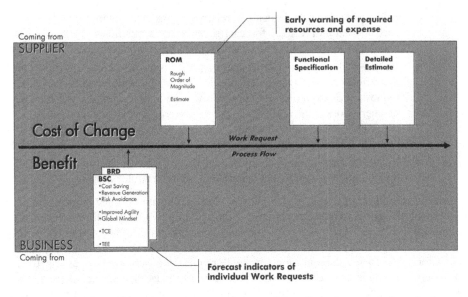

Fig. 1. Work request lifecycle: assessment of cost and benefit and accumulation of relevant information

2 Related Work

Bundling of work requests can be seen as a similar problem to bundling together services to be offered to users. Value Webs [1] represents an interesting approach to the service bundling problem, based on service ontologies. However the diversity and non-uniformity of the descriptions of work requests that we deal with in our case, make it difficult to apply such approaches. We argue here that using business rules to express dependencies between work requests strikes a nice tradeoff between being able to deal with complexity in the work request descriptions and ease of modeling from the point of view of a business user that needs to understand how work requests are dependent on one another.

Estimating the complexity of the software development effort necessary to fulfill a work request can be addressed through approaches such as the one described in [2]. In that category – and more relevantly to this work – Carr et. al [3] use a rule-based approach to describe dependencies among various artifact composing a software release, resulting in a very rich and expressive model. However, that line of work is complementary to what is presented here in that we look at a set of much simpler dependencies among work requests to begin with, with reserve to explore approaches inspired by and similar to [2, 3] as future work.

2.1 A Solution for Work Request Prioritization

The value proposition of our solution is threefold in that it proposes:

- Enhanced executive decision support capability for the best allocation of scarce development resources through timelier and more accurate delivery of forecast indicators relative to the net benefit. Examples of such indicators are total forecast revenue for a bundle of work requests and total forecast cost reductions for a bundle of work requests.

- Reduced development cost and shorter time to value through the identification of synergies, duplication and commonality in work requests. For example, our approach will be able to identify in a reliable manner work that falls into these discrete categories pertaining to potential duplication thereby highlighting areas of potential cost reduction.

- Reduced turn around time of delivery to trade customers through prioritization driven by net benefit and optimized release trains.

Our approach is represented in Figure 2. Customers submit work requests along with a set of business requirements. The submission also includes filling a balanced scorecard [4] based questionnaire. The questionnaire probes the submitters with questions on the likely impact that the work request will have on selected business indicators. Moreover, the questionnaire will request the submitter to assess their confidence in their answers. In order for the estimates not to be skewed, an adjustment is performed through the degree of *trust* that the system has in them. The trust factor is calculated by comparing the accuracy of the estimate along with the real values given by the suppliers after the work has been delivered.

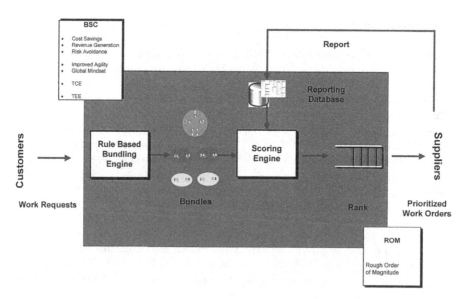

Fig. 2. Work requests bundling and prioritization design flow

The balanced scorecard (BSC) questionnaire is the input to a tool for decision support for the prioritization of work requests based on the HP labs Management by Business Objectives (MBO) technology [5]. The tool captures and makes explicit the process of eliciting the business value of the work requests through an assessment of the alignment of the expected outcome of fulfilling the work request with some explicitly stated business objectives. The work request is then sent to the corresponding suppliers (enablers, development teams) to give a ROM (Rough Order of Magnitude) estimate of the efforts needed to fulfill the request.

BSC and ROM are not just applied to work requests in isolation. Work requests are bundled together using a rule engine based on the Rete algorithm [6] where each dependency is represented through a rule and the work requests represent the facts that will be fed to the engine. In order to prioritize the bundles, a score based on the composition of the individual work requests balanced scorecard is calculated. The score is calculated depending on the type of the dependency. For example, if the type of the dependency represents a common code shared between the work requests, then this will result in a reduction in the efforts needed to deliver those work requests. Consequently, the metric associated with the cost savings will have a higher score. Requests with the highest score will be assigned the highest priority, so as to ensure the most effective use of the limited available development resources.

2.2 Bundling Work Requests

The questions of which work requests represent the highest net benefit for the business cannot be answered by considering individual work requests in isolation, but only by identifying and understanding the relationships between them. Our approach to this problem is to consider dependencies among work requests and bundle together work requests that satisfy criteria of *similarity*, *commonality* and *synergy*. Prioritization would then take place among bundles of work requests rather than individual work requests.

The similarity criterion applies to (partial) duplication of requests or requirements from different clients that result in re-use of development effort towards the satisfaction of multiple work requests. The commonality criterion applies to work requests having the same deadline, or being anyway part of a common strategic initiative. The synergy criterion applies when similar resources are required or the same code base needs to be modified.

For each type of work request dependency, we investigate how the prioritization methodology explained in the previous section can be extend to consider bundles of work requests rather than work requests in isolation. For instance, when similarity is identified between two or more work requests, we will have to adjust the ROM estimate to consider that the development effort will be reduced, since same or complementary products and code bases will have to be modified. On the other hand, when considering bundles of work requests linked together by commonality type of dependency (e.g. same deadline), the estimate on the delivery time will have to take into account the additional workload imposed onto the development teams with respect to what would be the case if considering each work request in isolation. Similar considerations apply to the business value estimate.

Construction of Bundles

The natural way of declarative describing dependencies among work requests is to use business rules. The mechanism we use to bundle work requests is illustrated through the example in Figure 3. There are three rules, (*r1*, *r2*, *r3*) corresponding to three types of dependencies (*dependency1*, *dependency2* and *dependency3*) and 11

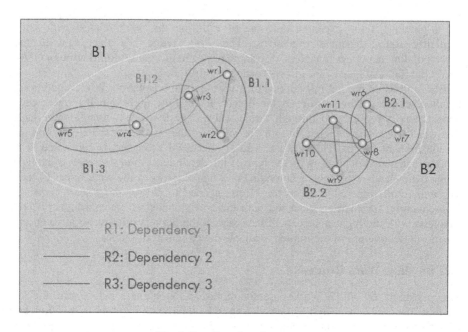

Fig. 3. Bundles of work requests

work requests (*wr1... wr11*). The head of each of the rules states the condition under which work requests would be bundled (example: same deadline). The body of the rule expresses the action to be undertaken when a dependency condition is satisfied. In most cases, this would result in bundling together the work requests, but a more detailed specification of the types of rules used is given in the next section. In the example in Figure 3, rule *r1* gets fired once and a *dependency1* is detected between work requests wr4 and wr3, creating bundle *B1.2*. Rule r2 gets fired 4 times, as *dependency2* is detected between (wr4, wr5), (wr6, wr7), (wr6, wr8) and (wr7, wr8). Since there are no dependencies between the groups (wr4, wr5) and (wr6, wr7, wr8), these two groups come to represent the two bundles *B1.3* and *B2.1*.

At the end of the first iteration of this process there are 5 bundles (*B1.1, B1.2, B1.3, B2.1, B2.2*). Each where each bundle is a graph composed of a set of work requests (nodes) sharing the same dependency (edges).

A work request, however, may have different dependencies with other work requests as is the case for *wr4* which has a *dependency1* with *wr3* and a *dependency2* with *wr5*. Since work requests sharing the same dependency are to be treated at the same time, dependent bundles (bundles sharing at least one work request) have also to be treated at the same time. Hence, this dependency is called an *indirect dependency*. Bundles *B1.1, B1.2* and *B1.3* are therefore grouped into bundle *B1* due to the indirect dependency. The same goes for *B2.1* and *B2.2* which are grouped into bundle *B2*.

Rules Declaration Through Decision Tables
We use *decision tables* [7] to encode the business rules. Decision tables are a user-friendly and easily maintainable notation for representing ECA (*Event-Condition-Action*) rules.

	Conditions				Action
	Same Customer	Same Deadline	Same Type	Type of Work Request	Bundle
Rule 1	Y	Y			dependency 1
Rule 2	Y		Y	Operational Efficiency	dependency 2

Fig. 4. Example of decision table: business view

Figure 4 shows an example of a decision table. The table comprises a set of *conditions* (colored in green) and a set of *actions* (colored in yellow) to perform. Conditions precede actions and are all placed in columns. A rule corresponds to a row in the table, and the intersection between a row and a column determines if a condition is part of the rule or not. When it is not, the cell is grayed out as shown in figure 4. A rule will fire if and only if *all* the corresponding conditions are met. For example, *Rule1* will fire if and only if both the "*Same Customer*" and "*Same Deadline*" conditions are "*true*". This will trigger the action "*Bundle*" to execute.

This simple table format makes it easy for business users (assumed to be less technical-savvy) to specify the conditions, actions, and rules in natural language. Users needn't worry about complex nested if then else statements or be confused by and/or when describing complex logic. Additionally, tables can be automatically checked to determine if rules are redundant or ambiguous. Designers can also decompose very complex logic into multiple tables, which invoke other tables as actions. The logic captured in the decision table is eventually translated into code by a developer so that it can later be executed by the rule engine (as it can be seen in figure 5, execution logic color-coded in orange).

20		RuleTable Bundling(VR a, VR b, BundleList list)				
21		CONDITION	CONDITION	CONDITION	CONDITION	ACTION
22	(descriptions)	a.getCustomer().equals(b. getCustomer())	a.getDeadline().equals(b.getDeadline())	a.getType().equals(b. getType())	a.getType().equals("$param")	list.add("$param",a.getId(),b. getId());
23		Same Customer	Same Deadline	Same Type	Type of Work Request	Bundle
24	Rule 1	Y	Y			dependency 1
25	Rule 2	Y		Y	Operational Efficiency	dependency 2

Fig. 5. Decision Table Example (technical and business view)

The collaborative process of interaction between the business user and the developer can be described through the following steps:

1. The business analyst retrieves a template decision table from a repository
2. The business analyst describes business logic in the table as seen in figure 4.
3. The decision table is handed to a developer, who maps the business language descriptions to the corresponding specification in the programming language of choice (Java™ in our prototypical example).
4. The developer hands back and reviews the modifications with the business analyst.

5. The business analyst can continue editing the rule rows as needed.
6. In parallel, the developer can develop test cases for the rules with the help of the business analysts, as these test cases can be used to verify rules and rule changes once the system is running.

Rules Execution and Bundles Creation

The rules expressed in decision tables as seen in the previous section are loaded in the working memory of the rule-based bundling system. Whenever new work requests are added as facts to the knowledge base, the rule engine schedules the rules for execution. The implementation of our prototype of the rule engine is based on the open source rule engine JBoss rules [8]. The consequence of executing ("firing") the rules is that the work requests are checked in pairs and whenever the set of conditions for a given rule are all satisfied, a dependency is created that links the pair of work requests together.

3   Prioritizing Bundles of Work Requests

Having defined bundles of work requests, we now go back to our original problem: given prioritization scores of individual work requests, what is the score associated to a bundle of work requests? Even though prioritization of single work requests is not the focus of this paper, it suffices to recall here that we assign scores to work requests based on business indicators as exemplified in figure 6.

For example, when considering an "*additional revenue earned*" indicator for a given work request, we associated a score to different intervals of values that might be selected by the business user of our tool. However, this method cannot be directly applied to bundles of work requests. Our approach is to therefore to look at each business indicator and analyze the effect of scoring more than one work request. The

		This is the only column that should be modified.				
ITEM	**SCORE** (Calculated)	**RATING**	**UNITS** (Standard)	**Total Allocation**	**Notes**
Revenue Generation Additional Revenue earned	9	>$1,000k Bronze <10 ▼	US$pa	20	1
Cost Reduction Cost Saving	18	$1,000k - $5,000k ▼	US$pa	20	2
Risk Avoidance Financial Penalties Avoided	7	$100k - $1,000k ▼	US$pa	10	3
Total	34			50	

Fig. 6. Financial perspective of the balanced scorecard tool to score individual work requests

effect could be cumulative as is the case for the "*additional revenue earned*" indicator: in this case, since a bundle of dependent work requests will be prioritized as all-or-nothing, the "*additional revenue earned*" value parameter for generating the score will be the sum of the scores for each work request in the bundle. In other cases, other scoring function (for instance the mean, median or maximum of the scores) will be taken into account. All of these criteria are naturally expressed through business rules and the reader can intuitively figure how these rules would be expressed.

On calculating scores for bundles of work requests, not only does our prototype take into account quantitative values as described in the example above, but it also considers intangibles such as efficiency and effectiveness of service delivery through cost savings, operational efficiency and increased customer and employee satisfaction. We devised a way of quantifying these intangible benefits of bundling together work requests and describe them too through business rules, as shown in the decision table in Figure 7. Following the Balanced Scorecard approach, business indicators are grouped into perspectives, such as "financial", "customer", etc. For each rule (dependency), the projected benefit along each perspective axis is expressed in percentage. For example, by applying rule 1, the resulting bundle will have an improvement of 25% on the financial perspective, 15% on the operational perspective and 5% at the total customer experience perspective.

	Actions			
	RuleTable Bundling[VR a, VR b, BundleList list]			
	ACTION	ACTION	ACTION	ACTION
	list.SetFinancialImprovement("$param");	list.SetOperationalImprovement("$param";	list.SetTCEImprovement("$param");	list.SetTEEImprovement("$param");
	Financial Improvement (%)	**Operational Improvement (%)**	**TCE Improvement (%)**	**TEE Improvement (%)**
Rule 1	25	15	5	
Rule 2	20	45		

Fig. 7. Business scorecard and perspective-driven improvements

To calculate the final score of a bundle, those improvements will be applied to the score at each perspective. For a bundle B, the final score is as follows:

$$FinalScore(B) = Score_{FinancialPerspective}(B) \quad * (1 + Improvement_{FinancialPerspective}(B))$$
+

$$Score_{OperationalPerspective}(B) * (1 + Improvement_{OperationalPerspective}(B)) \quad + \quad Score_{TCEPerspective}(B)$$
$$* (1 + Improvement_{TCEPerspective}(B)) \quad + \quad Score_{TEEPerspective}(B)$$
$$* (1 + Improvement_{TEEPerspective}(B))$$

4 Conclusion

In this paper, we have presented an approach for identifying synergies, duplication and commonality in work requests and grouping them together into bundles. We showed that bundling work requests could result into reduced development cost and shorter time to value. We also proposed a prioritization mechanism based on a balanced score card approach. The prioritization aims at favoring bundles of work requests generating the highest benefits.

Given that the nature of this paper is that of an experience report, we do not present a full validation section. However, we count as validation of our approach that the current version of the tool was deployed within the Managed Services organization of HP and had more than 400 users at the time we wrote.

We are working on extending the current version by enhancing the balanced scorecard tool through calculation of the alignment of with business objectives, and by providing a rule-based bundling tool that will capture all the other kinds of dependencies among work requests and adjust prioritization accordingly, according to the management by business objectives (MBO) methodology presented in [5].

Acknowledgment

The authors would like to thank Mark Mancq and Ian Lumb from HP Service for their precious help in defining the problems and shaping the requirements for this work.

References

[1] Akkermans, H., Baida, Z., Gordijn, J., Peiia, N., Laresgoiti, I.: Value Webs: using ontologies to bundle real-world services. IEEE Intelligent Systems, 19(4), 57–66 (2004)

[2] Basili, V., Briand, L., Condon, S., Kim, Y.M., Melo, W.L., Valen, J.D.: Understanding and predicting the process of software maintenance releases. In: ICSE 1996. Proc. 18th International Conference on Software Engineering, p. 464 (1996)

[3] Carr, M., Wagner, C.: A Study of Reasoning Processes in Software Maintenance Management. Journal of Information Technology and Management 3(1,2), 181–203 (2002)

[4] Kaplan, R.S., Norton, D.P.: The Balanced Scorecard: Translating Strategy into Action. Harvard Business School Press (1996)

[5] Bartolini, C., Sallé, M., Trastour, D.: IT service management driven by business objectives – an application to incident management. In: NOMS 2006. Proc. IEEE/IFIP Network Operation and Management Symposium, IEEE Computer Society Press, Los Alamitos (2006)

[6] Forgy, C.: Rete: A Fast Algorithm for the Many Patterns/Many Objects Match Problem'. Artificial Intelligence Journal 19, 17–37 (1982)

[7] Pollack, S.L., Hicks, H.T., Harrison, W.J.: Decision tables: theory and practice. Wiley-Interscience, Chichester (1971)

[8] JBoss Rules (URL last checked July 2007), www.jboss.com/products/rules

A Generic Module System for Web Rule Languages: Divide and Rule

Uwe Aßmann[1], Sacha Berger[2], François Bry[2], Tim Furche[2], Jakob Henriksson[1], and Paula-Lavinia Pătrânjan[2]

[1] Technische Universität Dresden,
Nöthnitzer Str. 46, D-01187 Dresden, Germany
http://www-st.inf.tu-dresden.de/
[2] Institute for Informatics, University of Munich,
Oettingenstr. 67, D-80538 München, Germany
http://www.pms.ifi.lmu.de/

Abstract. An essential feature in practically usable programming languages is the ability to encapsulate functionality in reusable modules. Modules make large scale projects tractable by humans. For Web and Semantic Web programming, many rule-based languages, e.g. XSLT, CSS, Xcerpt, SWRL, SPARQL, and RIF Core, have evolved or are currently evolving. Rules are easy to comprehend and specify, even for non-technical users, e.g. business managers, hence easing the contributions to the Web. Unfortunately, those contributions are arguably doomed to exist in isolation as most rule languages are conceived without modularity, hence without an easy mechanism for integration and reuse. In this paper a generic module system applicable to many rule languages is presented. We demonstrate and apply our generic module system to a Datalog-like rule language, close in spirit to RIF Core. The language is gently introduced along the EU-Rent use case. Using the Reuseware Composition Framework, the module system for a concrete language can be achieved almost for free, if it adheres to the formal notions introduced in this paper.

1 Introduction

Modules are software units that group together parts of different programs and knowledge (or data structures) and that usually serve a specific purpose. For any practically applied programming language modules are an essential feature, as they make large scale projects tractable by *humans*. Modules not only facilitate the integration of existing applications and software into complex projects, they also offer a flexible solution to application development where part of the application logic is subject to change.

The work presented in this article advocates for the need and advantages of modularizing rule-based languages. The rule-based paradigm offers elegant and flexible means of application programming at a high-level of abstraction. During the last couple of years, a considerable number of initiatives have focused on using rules on the Web (e.g. RuleML, R2ML, XSLT, RIF Core etc.).

A. Paschke and Y. Biletskiy (Eds.): RuleML 2007, LNCS 4824, pp. 63–77, 2007.

However, unless these rule languages are conceived with proper support for encoding knowledge using modular designs, their contributions to the Web are arguably doomed to exist in isolation, hence with no easy mechanism for integration of valuable information. Many of the existing rule languages on the Web do not provide support for such modular designs. The rationale behind this is the focus on the initially more crucial concerns relating to the development of the languages.

The main difference between the module system presented here and previous module system work is, that our approach focuses on genericity. That is, the approach is applicable to many rule languages with only small adaptation to syntactic and semantic properties of the language to be supported with constructs for modularization. The concept is based on rewriting modular programs into semantically equivalent non-modular programs, hence not forcing the evaluation of the hosting language to be adapted to some operational module semantics. This approach not only enables reuseability for current languages, it is arguably also applicable to forthcoming rule languages. The presented module system hence arguably enables reuseability in two aspects—it not only supports users of languages to design programs in a modular fashion, it also encourages tool and language architects to augment their rule languages by reusing the abstract module system. Last but not least: the module system is <u>easy</u> in several aspects. First, it is very easy for language developers to apply due to the employment of reduction semantics of a given modularized rule language to its un-modular predecessor. Second, the reduction semantics is kept simple. There is just one operator, yet it is sufficient to model a quite rich modular landscape including visibility, scoped module use, parametric modules etc., though we disallow recursive modules, for reasons of simplicity. Third, the implementation of the abstract module system can be achieved using existing language composition tools, for example, Reuseware.[1] A concrete modularized language is achieved by mere instantiation of the abstract implementation, making the implementation of the abstract module system fast and easy.

The main contribution of our work is a conceptual framework for modules in rule languages. We identify requirements defining a class of rule languages that can be extended with the proposed module system. The conceptual framework provides abstract language constructs for a module system, which are to extend the syntax of the hosting rule language. The principles of modularization and the main language constructs are gently introduced using Datalog as an example rule language.

We have evaluated the ideas presented in this paper in a companion work [2] that shows how to add modules to the rule-based Web query language Xcerpt. In particular, it shows that the ideas can be applied even for highly expressive, fairly complex rule languages such as Xcerpt.

The paper is structured as follows. Section 2 introduces the rule language Datalog and demonstrates the usability of modularization of its rules via a use-case. Section 3 describes and motivates the main requirements of any rule language in order for it to be applicable to the underlying module framework. Section 4 describes the underlying module algebra for the framework. Section 5 compares our work to other module approach and Section 6 concludes the paper with some final remarks.

---

[1] http://reuseware.org

2   Module Extension by Example

Rules in a particular language are usually specified in a rule set, a finite set of (possibly) related rules. Recall from Section 1 that we focus our work on so-called deductive rules. A deductive rule consists of a head and a body and has the following form:

$$head :\text{-} body$$

where *body* and *head* are finite sequences of atomic formulas of the rule language. The formulas of each set are composed using (language specific) logical connectives. The implication operator :- binds the two rule parts together. Rules are read "if the body holds, then the head also holds". Usually, rule parts share variables: the variable substitutions obtained by evaluating the *body* are used to construct new data in the *head*.

In the following we extend a concrete rule language with modules. By following our proposed approach, which is formalized in Section 4, one can afford to abstract away from a particular data model or specific capabilities supported by a rule language and - most important - not to change the semantics of the language.

2.1   Datalog as an Example Rule Language

In the introductory part we mentioned Datalog as an example rule language based on deductive rules. Datalog is a well-known database query language often used for demonstrating different kinds of research results (e.g. query optimization). Datalog-like languages have been successfully employed, e.g. for Web data extraction in Lixto[2]. The strong similarities between Datalog and RIF Core suggest that the ideas followed for modularizing Datalog could be also applied to RIF Core.

In Datalog, the *head* is usually considered to be a singleton set and the *body* a finite set of Datalog atoms connected via conjunction ($\wedge$). A Datalog atom is an n-ary predicate of the form $p(a_1, \ldots, a_n)$, where $a_i (1 \leq i \leq n)$ are constant symbols or variables. As such, a Datalog rule may take the following form:

```
_____ A Simple Datalog Rule _____
gold-customer(Cust)  :- rentals(Cust,Nr,2006), Nr > 10.
```

to be understood as defining that *Cust* is a gold customer if the number Nr of car rentals *Cust* made in 2006 is greater than 10.

Rules are associated a semantics, which is specific to each rule language and cannot be described in general terms. For the case of Datalog, semantics is given by definition of a least Herbrand model of a rule-set. We don't go into details regarding the semantics, since our work on modularizing Datalog preserves the language's semantics.

Datalog, and more general deductive rules, infer new knowledge (called intensional knowledge) from existing, explicitly given knowledge. As already recognized [11], there is a need to 'limit' the amount of data used in performing inference on the Web – a big and open source of knowledge. Thus, the notion of *scoped inference* has emerged. The idea is to perform inference within a scope of explicitly given knowledge sources.

---

[2] Data extraction with Lixto,
http://www.lixto.com/li/liview/action/display/frmLiID/12/

One elegant solution for implementing scoped inference is to use *modules* for separating the knowledge. In such a case, the inference is performed within a module. Since inference is essential on the Web and, thus, modules for rule languages such as Datalog, let's see how we could modularize Datalog!

2.2 Module Extension for Datalog

This section gives a light introduction to modularizing a rule language such as Datalog that should ease the understanding of the formal operators proposed in Section 3. We consider as framework for our examples the EU-Rent[3] case study, a specification of business requirements for a fictitious car rental company. Initially developed by Model Systems Ltd., EU-Rent is promoted by the business rules community as a scenario for demonstrating capabilities of business rules products.

The concrete scenario we use for showing advantages of introducing modularization in rule languages is similar to the use case for rule interchange 'Managing Inter-Organizational Business Policies and Practices', published by the W3C Rule Interchange Format Working Group in the 2nd W3C Working Draft of 'RIF Use Cases and Requirements'. The car rental company EU-Rent operates in different EU countries. Thus, the company needs to comply with existing EU regulations and each of its branches needs also to comply with the regulations of the country it operates in.

The EU-Rent company heavily uses rule-based technologies for conducting its business. This was a straightforward choice of technology from the IT landscape, since rule languages are more than suitable for implementing company's policies. Moreover, EU regulations are also given as (business) rules.

Different sets of rules come into play for most of the company's rental services, such as advance reservations for car rentals or maintenance of cars at an EU-Rent branch. A set of rules implement, as touched on above, the company's policies (e.g. that *the lowest price offered for a rental must be honored*). These rules are used by each of the EU-Rent branches. Another set of rules implements the policies used at a particular EU-Rent branch, as they are free to adapt their business to the requirements of the market they address (of course, as long as they remain in conformance with the EU-Rent company-level rules). As is the case for EU regulations, EU-Rent branches might need to comply with existing national regulations—an extra set of rules to be considered.

We have illustrated so far a typical scenario for data integration. The sets of rules our EU-Rent company needs to integrate for its services may be stored locally at each branch or in a distributed manner (e.g. company level rules are stored only at EU-Rent Head Quarter and EU and national rule stores exist on different servers for the corresponding regulations). Rules might change at the company level and regulations might also change both at EU and at national level. So as to avoid the propagation of such changes every time they occur, a distributed and modularized approach to the EU-Rent implementation would be a suitable solution.

An architectural overview of the scenario described so far is given above. The overview sketches a possible modularization of rules employed by the EU-Rent company. Modules are represented here as boxes. An example EU-Rent branch, the Paris

---

[3] EU-Rent case study, http://www.businessrulesgroup.org/egsbrg.shtml

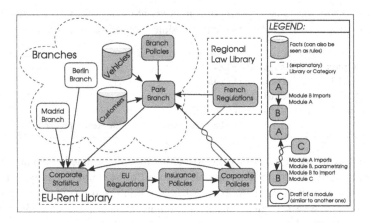

Fig. 1. EU-Rent Use Case: Module Structure

branch, subdivides its vehicle and customer data as well as its policies into different modules, hence separating concerns for gaining clarity and ease maintenance. The module `ParisBranch` *imports* the defined modules and, thus, uses their rules together with those defined in `ParisBranch`. Such module dependencies (i.e., *imports*) are indicated by arrows between boxes.

Modules can be imported and their rules can be defined as *private* or *public*. Private rules are not visible, i.e. the knowledge inferred by such rules can not be accessed directly in modules importing them. A statement `private import M2` in a module M1 makes all rules of M2 private within M1 and thus invisible to the modules importing M1. Rules defined as public can be used directly in modules importing them. By importing a module M as public makes its (public) rules visible in all modules importing M.

In the following we specify an excerpt of the module `ParisBranch` in Datalog:

```
ParisBranch
import private Vehicles
import private Customers
import private BranchPolicies
import private CorporatePolicies ( regional-law = "FrenchRegulations" )
public vehicle(X) :- voiture(X).
public vehicle(X) :- bicyclette(X).
```

The `CorporatePolicies` module, imported in the module example above, can be considered as the core of the depicted architecture. It aggregates various (domain specific) modules like `EURegulations` and `InsurancePolicies`, which implement relevant European insurance regulations. Branches import the `CorporatePolicies` module and comply thus to corporate policy. By design, the company branches use the corporate policies module as a common access point also to local regulations. As the previous example shows, the Paris branch (and other EU-Rent branches too) parameterizes the module `CorporatePolicies` using as parameter the local regulations the branch needs to comply to. More concretely, the parameter `regional-law` is instantiated with the name of the module implementing the specific regulations, in this case FrenchRegulations. We define the module CorporatePolicies as follows:

```
───────────────────────── CorporatePolicies ─────────────
declare parameter regional-law
import public InsurancePolicies
import private EURegulations
import public parameter regional-law
...                        // rules defining the Eu-Rent policies
```

In the following we turn our attention to another module depicted in our architecture, viz. CorporateStatistics. This module imports CorporatePolicies, but also all branch modules (such as ParisBranch, MadridBranch, and BerlinBranch), a task doomed to produce vast naming clashes of symbols defined in the imported modules. To overcome the problem, a *qualified import* is to be used: Imported modules get local names that are further used to disambiguate the symbols. Thus, one can smoothly use knowledge inferred by different, imported rules with same heads.

The following example shows how Datalog can be extended with a qualified import. To have an overview over the status of EU-Rent vehicles, the notion of an old vehicle is defined differently for the different branches. The modules ParisBranch, Munich-Branch, and MadrdidBranch get the local names m1, m2, and m3, respectively. The local names are associated with the module using @.

```
───────────────────────── CorporateStatistics ─────────────
import private CorporatePolicies
import ParisBranch @ m1
import MunichBranch @ m2
import MadridBranch @ m3
private old-vehicle(V) :- m1.vehicle(V), manufactured(V,Y), Y < 1990.
private old-vehicle(V) :- m2.vehicle(V), manufactured(V,Y), Y < 2000.
private old-vehicle(V) :- m3.vehicle(V), manufactured(V,Y), Y < 1995.
```

The simple module-based architecture and the given module examples show a couple of advantages of such a module-based approach. We have already touched on the *separation of concerns* when describing the modules of Paris branch as having each a well-defined purpose. Modules such as EURegulations and FrenchRegulations can be *reused* by other applications too (not only by new established EU-Rent branches, but also by other companies). It could even be published by government agencies on the Web. A module-based implementation is much more flexible and less error-prone than one without modules; this eases considerably the *extensibility* of the implementation.

3   Framework for Rule Language Module Systems

In Section 2 we saw an example of how it is possible to modularize rule-sets—in this case for Datalog—and that it is important to ensure separation (or encapsulation) of the different modules. We choose to enforce this separation statically, i.e., at compile time, due to our desire to reuse—rather than extend—existing rule engines that do not have an understanding of modules. An added advantage is a clean and simple semantics based on concepts already familiar to the users of the supported rule language.

We are interested in extending what is done for Datalog in Section 2 to a general framework for rule languages. The notion of modules is arguably important not only for Datalog, but for any rule language lacking such an important concept.

However, our reduction semantics for module operators poses some requirements to the expressiveness of a rule language: To describe these requirements, we first introduce in the following a few notions and assumptions on rule languages that give us formal means to talk about a rule language in general. Second, we establish the single requirement we ask from a rule language to be amenable to our module framework: the provision of rule dependency (or rule chaining).

In the next section, we then use these notions to describe the single operator needed to formally define our approach to modules for rule languages. We show that, if the rule language supports (database) views, that single operator suffices to obtain a powerful, yet simple to understand and realize module semantics. Even in absence of (database) views, we can obtain the same result (from the perspective of the module system's semantics) by adding two additional operators.

Rule languages. For the purpose of this work, we can take a very abstract view of a rule language: A rule language is any language where 1. <u>programs</u> are finite sequences of rules of the language; 2. <u>rules</u> consist of one head and one body, where the body and the head are finite sequences of rule parts; 3. <u>rule parts</u>[4] are, for the purpose of this work, arbitrary. They are not further structured though they may in fact be negative literals or complex structures. Body rule parts can be understood as a kind of condition and head rule parts as a kind of result, action, or other consequence of the rule. We use indices to identify a rule within a program as well as head or body parts within a rule. Note, that we do not pose any limitations to the shape of the body parts or the connectors used (conjunction, disjunction, etc.).

Rule dependency. Surprisingly, we care very little about the actual shape of rule parts, let alone their semantics. The only critical requirement needed by our framework is that the rule language has a concept of <u>rule chaining</u> or <u>rule dependency</u>. That is, one rule may depend on another for proper processing.

Definition 1 (Rule dependency). *With a program P, a (necessarily finite) relation $\Delta \subset \mathbb{N}^2 \times \mathbb{N}^2$ can be associated such that $(r_1, b, r_2, h) \in \Delta$ iff the condition expressed by the b-th body part of rule r_1 is dependent on results of the h-th head part of the r_2-th rule in P (such as derived data, actions taken, state changes, or triggered events), i.e., it may be affected by that head part's evaluation.*

Controlling rule (or rule part) dependency can take different forms in different rule languages: in Datalog, predicate symbol provide one (easy) means to partition the dependency space; in XSLT, modes can serve a similar purpose; in Xcerpt, the labels of root terms; etc. However, the realisation of the dependency relation is left to the rule language. We assume merely that it can be manipulated arbitrarily, though the module

---

[4] We refrain from calling rule parts literals, as they may be, e.g., entire formulas or other constructs such as actions that are not always considered logical literals.

system never introduces cycles in the dependency relation if they do not already exist. Thus the rule language does not <u>need</u> to support recursive rules for the reduction semantics to be applicable, however, if present, recursive rules pose no challenge. We do, however, assume that the dependencies between <u>modules</u> are non-recursive.

Observe, that these rewritings may in fact affect the constants used in the program. However, constants are manipulated in such a way that, for each module, there is an isomorphism between the rewritten program and the original program. Assuming a generic [1] rule language, this renaming has no effect on the semantics of the program.

The module extension framework requires the ability to express an arbitrary (acyclic) dependency relation, however poses no restrictions on the shape of Δ for a module-free program. Indeed, for any module-free rule program P the unrestricted relation with all pairs of head parts and body parts forms a perfectly acceptable Δ relation on P.

We only require the ability to express <u>acyclic</u> dependency relations as the discussed module algebra does not allow cycles in the module composition for simplicity's sake.

Most rule languages that allow some form of rule chaining, e.g., datalog, SWRL, SQL, Xcerpt, R_2G_2, easily fulfill this requirement. However, it precludes rule languages such as CSS where all rules operate on the same input and no rule chaining is possible. Interestingly, though CSS already provides its own module concept, that module concept provides no information hiding, the central aim of our approach: Rules from all imported modules are merged into one sequence of rules and all applied to the input data, only precedence, not applicability, may be affected by the structuring in modules.

Reduction semantics. Why do we impose the requirement to express arbitrary dependency relations on a rule language to be amenable to our module framework? The reason is that we aim for a reduction semantics where all the additions introduced by the module framework are reduced to expressions of the original language. To achieve this we need a certain expressiveness which is ensured by this requirement.

Consider, for instance, the Module "CorporateStatistics" in Section 2. Let's focus only on old-vehicle and the three vehicle predicates from the three local branches. Using the semantics defined in the following section, we obtain a program containing also all rules from the included modules plus a dependency relation that enforces that only certain body parts of old-vehicle depend on the vehicle definition from each of the local branches. This dependency relation can be <u>realized</u> in datalog, e.g., by properly rewriting the predicate symbols through prefixing predicates from each of the qualified modules with a unique prefix.

4   Module System Algebra

Remember, that the main aim of this work is to allow a rule program to be divided into <u>conceptually independent collections</u> (modules) of rules with well-defined interfaces between these collections.

For this purpose, we introduce in Section 4.1 a formal notion of a collection of rules, called "module", and its public interface, i.e., that subset of rules that constitutes the (public) interface of the module. Building on this definition, we introduce an algebra (consisting in a single operator) for composing modules in Section 4.2 together with a reduction semantics, i.e., a means of reducing programs containing such operators to module-free programs.

Operators by example. Before we turn to the formal definitions, let's again consider the EU-Rent use case from Section 2, focusing on the three modules "CorporateStatistics", "CorporatePolicies", and "ParisBranch".

We can define all the import parts of these modules using the module algebra introduced in the following. We use $A \times B$ for indicating that a module B is imported into module A and A inherits the public interface of B (cf. import public), $A \bowtie B$ to indicate private import (cf. import private), and $A \bowtie_S B$ for scoped import (cf. import ... @) where S are pointers to all rule parts addressing a specific module. Note, that public and private import can actually be reduced to the scoped import if the language provides views.

Using these operators we can build formal module composition expressions corresponding to the surface syntax from Section 2 as follows:

$$\mathsf{CorporatePolicies}' = (\mathsf{CorporatePolicies} \times \mathsf{InsurancePolicies}) \bowtie \mathsf{EURegulations}$$
$$\mathsf{CorporatePolicies}'_{french} = \mathsf{CorporatePolicies}' \times \mathsf{FrenchRegulations}$$
$$\mathsf{ParisBranch}' = (((\mathsf{ParisBranch} \bowtie \mathsf{Vehicles}) \bowtie \mathsf{Customers}) \bowtie \mathsf{BranchPolicies})$$
$$\bowtie \mathsf{CorporatePolicies}'_{french}$$
$$\mathsf{CorporateStatistics}' = (((\mathsf{CorporateStatistics} \bowtie \mathsf{CorporatePolicies}') \bowtie_{(1,1)} \mathsf{ParisBranch}')$$
$$\bowtie_{(2,1)} \mathsf{MunichBranch}') \bowtie_{(3,1)} \mathsf{MadridBranch}'$$
$$\mathsf{MunichBranch}' = \ldots$$

Thus, given a set of basic modules, each import statement is translated into a module algebra expression that creates a new module, viz. the semantics of the import statement. Unsurprisingly, parameterized modules lead to multiple "versions" of similar module composition expressions that only differ in instantiations of the parameters.

4.1 Defining Modules

We use <u>module identifiers</u> as means to refer to modules, e.g., when importing modules. Some means of resolving module identifiers to modules (stored, e.g., in files, in a database, or on the Web) is assumed, but not further detailed here.

Definition 2 (Module). *A module M is a triple* $(R_{\mathrm{PRIV}}, R_{\mathrm{PUB}}, \Delta) \subset \mathscr{R} \times \mathscr{R} \times \mathbb{N}^4$ *where $\mathscr{R}$ is the set of all finite sequences over the set of permissible rules for a given rule language. We call R_{PRIV} the private, R_{PUB} the public rules of M, and Δ the dependency relation for M. For the purpose of numbering rules, we consider $R = R_{\mathrm{PRIV}} \diamond R_{\mathrm{PUB}}$[5] the sequence of all rules in M.*

We call a module's public rules its "interface": When importing this module, only these rules become accessible to the importing module in the sense that rule bodies in the importing module may depend also on the public rules of the imported module but not

---

[5] $\diamond$ denotes <u>sequence concatenation</u>, i.e., $s_1 \diamond s_2$ returns the sequence starting with the elements of s_1 followed by the elements of s_2, preserving the order of each sequence.

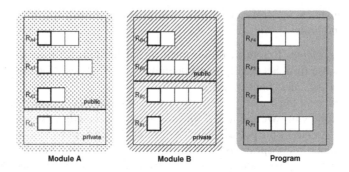

Fig. 2. Program and two defined modules without imports

on its private rules. Though the module composition discussed in this section does not rely on any further information about a module, a module should be accompanied by additional information for its <u>users</u>: documentation about the purpose of the module, what to expect from the modules interface, what other modules it relies on, etc.

Figure 2 shows an exemplary configuration of a program (which is just a module where R_{PUB} is empty) together with two modules A and B. Where the program consists of a single sequence of (private) rules, the rules of each of the modules are partitioned into private and public rules. The allowed dependency relation Δ is represented in the following way: All body parts in each of the areas ⬚, ▨, and ■ are depending on all head parts in the same area and no other head parts. No access or import of modules takes place, thus no inter-module dependencies exists in Δ between rule parts from one of the modules with each other or with the (main) program.

Notice, that for the dependency within a module the partitioning in private and public plays <u>no role whatsoever</u>. Body parts in private rules may access head parts from public rules and vice versa.

4.2 Module Composition

Module composition operators allow the (principled, i.e., via their public interface) definition of inter-module dependencies. Our module algebra (in contrast to previous approaches) needs only a single fundamental module composition operator. Further operators can then be constructed from a combination of that fundamental operator and (standard database) views. However, we also discuss an immediate definition of these operators for languages where the view based approach is not applicable or desirable.

Scoped import. The fundamental module composition operator is called the <u>scoped import operator</u>. Its name stems from the fact that it allows to specify not only which module to import but also which of the rules of the importing module or program may be affected by that import and thus the scope of a module's import.

Informally, the scoped import of B in A uses two modules A and B and a set S of body parts from A that form the scope of the module import. It combines the rules from B with the rules from A and extends the dependency relation from all body parts in S to all public rules from B. No other dependencies between rules from A and B are established.

To illustrate this consider again the configuration from Figure 2. Assume that we import (1) module A into B with a scope limited to body part 3 of rule R_{B2} and that (2) we import the result into the main program limiting the scope to body parts 2 and 3 of rule R_{P1}. Third, we import module A also directly into the main program with scope body part 1 of R_{P1}.

This can be compactly expressed by the following module composition expression:

$$\left(P \bowtie_{(1,2),(1,3)} \left(B \bowtie_{(2,3)} A\right)\right) \bowtie_{(1,1)} A$$

As usual, such expressions are best read inside out: We import the result of importing A into B with scope $\{(2,3)\}$ into P with scope $\{(1,2),(1,3)\}$ and then also import A with scope $\{(1,1)\}$. The result of this expression is shown in Figure 3.

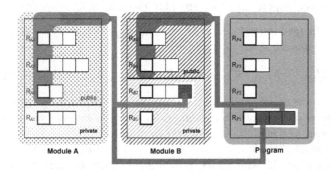

Fig. 3. Scoped import of (1) module A into body part 3 of rule R_{B2} and into body part 1 of rule R_{P1} and (2) of (the expanded) module B into body part 2 and 3 of rule R_{P1}. into the main program

Formally, we first introduce the concept of (module) scope.

Definition 3 (Scope). *Let $M = (R_{\mathrm{PRIV}}, R_{\mathrm{PUB}}, \Delta)$ be a module (or program if R_{PUB} is the empty sequence). Then a set of body parts from M is called a scope in M. More precisely, a scope S in M is a set of pairs from $\mathbb{N}^2$ such that each pair identifies one rule and one body part within that rule.*

For instance, the scope $\{(1,2),(1,3),(4,2)\}$ comprises for program P from Figure 2 the body parts 2 and 3 of rule 1 and the body part 2 of rule 4.

Second, we need a notation for adjusting a given dependency relation when adding rules. It turns out, a single operation (slide) suffices for our purposes:

Definition 4 (Dependency slide). *Given a dependency relation Δ, slide computes a new dependency relation by sliding all rules in the slide window $W = [s+1, s+length+1]$ in such a way that the slide window afterwards starts at $s_{new} + 1$:*

$$slide(\Delta, s, length, s_{new}) = \{(r_1', b, r_2', h) : (r_1, b, r_2, h) \in \Delta$$

$$\wedge r_1' = \begin{cases} s_{new} + 1 + (r_1 - s) & \text{if } r_1 \in W \\ r_1 & \text{otherwise} \end{cases} \wedge r_2' = \begin{cases} s_{new} + 1 + (r_2 - s) & \text{if } r_2 \in W \\ r_1 & \text{otherwise} \end{cases} \}$$

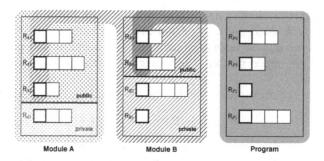

Fig. 4. Private import of A into B and B into the main program

With this, a scoped import becomes a straightforward module composition:

Definition 5 (Scoped import $\bowtie$). *Let* $M' = (R'_{\text{PRIV}}, R'_{\text{PUB}}, \Delta')$ *and* $M'' = (R''_{\text{PRIV}}, R''_{\text{PUB}}, \Delta'')$ *be two modules and S a scope in M'. Then*

$$M' \bowtie_S M'' := (R_{\text{PRIV}} = R'_{\text{PRIV}} \diamond R''_{\text{PRIV}} \diamond R''_{\text{PUB}}, R_{\text{PUB}} = R'_{\text{PUB}}, \Delta'_{\text{slided}} \cup \Delta''_{\text{slided}} \cup \Delta_{\text{inter}}), \text{ where}$$

- $\Delta'_{\text{slided}} = \text{slide}(\Delta', |R'_{\text{PRIV}}|, |R'_{\text{PUB}}|, |R_{\text{PRIV}}|)$ *is the dependency relation of the* importing *module M' with the* public *rules slided to the very end of the rule sequence of the new module (i.e., after the rules from M''),*
- $\Delta''_{\text{slided}} = \text{slide}(\Delta'', 1, |R''_{\text{PRIV}}| + |R''_{\text{PUB}}|, |R'_{\text{PRIV}}|)$ *is the dependency relation of the* imported *module M' with all its rules slided between the private and the public rules of the importing module (they have to be "between" because they are part of the private rules of the new module),*
- $\Delta_{\text{inter}} = \{(r_1, b, r_2, h) : (r_1, b) \in S \wedge \exists \text{ a rule in } R_{\text{PRIV}} \text{ with index } r_2 \text{ and head part } h : r_2 > |R'_{\text{PRIV}}| + |R''_{\text{PRIV}}|\}$ *the inter-dependencies between rules from the importing and rules from the imported module. We simply allow each body part in the scope to depend on all the public rules of the imported module. This suffices for our purpose, but we could also choose a more precise dependency relation (e.g., by testing whether a body part can at all match with a head part).*

Note, that the main difficulty in this definition is the slightly tedious management of the dependency relation when the sequence of rules changes. We, nevertheless, chose an explicit sequence notation for rule programs to emphasize that this approach is entirely capable of handling languages where the order of rules affects the semantics or evaluation and thus should be preserved.

Public and private import. Two more import operators suffice to make our module algebra sufficiently expressive to formalize the module system discussed in Section 2 as well as module systems for languages such as XQuery or Xcerpt.

In fact, if the language provides a (database) view concept, the single scoped import operator suffices as the two remaining ones can be defined using views on top of scoped imports. Before introducing these rewritings, we briefly introduce the public and private import operator. Formal definitions are omitted for conciseness reasons, but can be fairly straightforwardly derived from the definition for the scoped import.

All three operators hide information resulting from private rules in a module, however the public information is made accessible in different ways by each of the operators: The scoped import operator makes the information from the public interface of B accessible only to explicitly marked rules. The private and public import operators, in contrast, make all information from the public interface of B accessible to all rules of A. They differ only w.r.t. cascading module import, i.e., when a module A that imports another modules B is itself imported. In that case the public import operator ($\times$) makes the public interface of B part of the public interface of A, whereas the private import operator ($\bowtie$) keeps the import of B hidden.

Figure 4 shows the effect of the private import operator on the configuration from Figure 2 using the module composition expression P $\bowtie$ (B $\bowtie$ A): Module B imports module A privately and the main program imports module B privately. In both cases, the immediate effect is the same: The body parts of B get access to the head parts in A's public rules and the body parts of the main program P get access to the head parts in B's public rules. The import of A into B is hidden entirely from the main program. This contrasts to the public import in the expression P $\times$ (B $\times$ A). There the main program's body parts also depend on the head parts in A's (and not only B's) public rules.

Operator Rewriting. As stated above, we can express both public and private import using additional views (i.e., deductive rule) plus a scoped import, if the rule language provides view.

Theorem 6 (Rewriting $\times$). *Let* $M' = (R'_{\text{PRIV}}, R'_{\text{PUB}}, \Delta'), M'' = (R''_{\text{PRIV}}, R''_{\text{PUB}}, \Delta'')$ *be modules and* $M = (R_{\text{PRIV}}, R_{\text{PUB}}, \Delta) = M' \times M''$. *Then,* $M^* = (R'_{\text{PRIV}}, R'_{\text{PUB}} \diamond R, \Delta') \bowtie_S M''$ *is (up to the helper predicate R) equivalent to M if*

$$R = [h\!:\!-\,h : h \text{ is a head part in } R''_{\text{PUB}}]$$
$$S = \{(i,1) : |R'_{\text{PRIV}}| + |R'_{\text{PUB}}| < i \leq |R'_{\text{PRIV}}| + |R'_{\text{PUB}}| + |R|\}.$$

The gist is the introduction of "bridging" rules that are dependent on no rule in the existing module but whose body parts are the scope of the import of M''.

Note, that we use one rule for each public head part of the imported module. If the language provides for body parts that match any head part (e.g., by allowing higher-order variables or treating predicate symbols like constant symbols like in RDF), this can be reduced to a single additional rule. In both cases, however, the rewriting is linear in the size of the original modules.

For private import, an analogous corollary holds (where the "bridge" rules are placed into the private rules of the module and the dependency relation properly adapted).

5 Related Work

Despite the apparent lack of modules in many Web rule languages, module extensions for logic programming and other rule languages have been considered for a long time in research. We believe, that one of the reasons that they are still not part of the "standard repertoire" of a rule language lies in the complexity of many previous approaches.

For space reasons, we can only highlight a few selected approaches. First, in logic programming module extensions for Prolog and similar languages have fairly early been considered, cf., e.g., [4,13,6]. Miller [12] proposes a module extension for Prolog that includes parameterized modules similar in style to those as discussed in Section 2 and is the first to place a clear emphasis on strict (i.e., not to overcome) information hiding. In contrast to our approach, the proposed semantics requires an extension of standard logic programming (with implication in goals and rule bodies).

A reduction semantics, as used in this paper, is proposed in [10], though extra logical run-time support predicates are provided to allow module handling at run-time. However, the approach lacks support for module parameters and a clear semantics (most notable in the distinction between import and merge operation).

The most comprehensive treatment of modules in logic programming is presented in [5]. The proposed algebra is reminiscent of prior approaches [3] for first-order logic modules in algebraic specification formalisms (such as [14]). It shares a powerful expressiveness (far beyond our approach) and beautiful algebraic properties thanks to a full set of operators such as union, intersection, encapsulation, etc. The price, however, is that these approaches are far more complex. We believe, that a single well-designed union (or combination) operation together with a strong reliance on views as an established and well-understood mechanism in rule languages is not only easier to grasp but also easier to realize. E.g., intersection and renaming operations as proposed in [3] can be handled by our module algebra through a combination of scoped imports and views. More recently, modules have also been considered in the context of distributed evaluation [7] which is beyond the scope of our paper.

6   Conclusion

This work started with the ambition to abstract from relatively similar module systems realizable for different rule languages. In accordance with this ambition we defined a framework with clearly identifiable requirements put on rule languages to be usable in the framework. An important point to make, and a great advantage with such a framework, is that not only current rule language lacking constructs for modular design can be addressed, but also other rule languages which are yet to be designed and developed.

For the framework to be viable in practice, the concrete module implementation for a specific rule language falls back to the semantics and tools of the underlying rule language. Thus, the underlying tools can be used unmodified. Other techniques exist that take a similar 'reductive' approach to program manipulation, for example the composition framework Reuseware [9,8]. Due to these similarities, it is possible to implement the rewriting mechanism assumed in the module framework using the approach and tools provided by the composition framework Reuseware. As such, the practical implementation needed for different rule languages to support modular development can rely on existing work.

We have already implemented module systems for several languages based on the ideas presented here, in particular for the logic programming-style query language Xcerpt, for a rule-based grammar language R2G2, and for a Datalog fragment as discussed here. In all cases, the presented module system proved expressive enough to yield a useful yet easy to use language extension with minimal implementation effort. Application of the framework to further languages are under work.

Acknowledgments

This research has been funded by the European Commission and by the Swiss Federal Office for Education and Science within the 6th Framework Programme project REWERSE number 506779 (cf. http://rewerse.net).

References

1. Abiteboul, S., Hull, R., Vianu, V.: Foundations of Databases. Addison-Wesley Publishing Co, Boston, MA, USA (1995)
2. Aßmann, U., Berger, S., Bry, F., Furche, T., Henriksson, J., Johannes, J.: Modular web queries—from rules to stores (submitted for publication, 2007)
3. Bergstra, J.A., Heering, J., Klint, P.: Module algebra. Journal of the ACM 37(2), 335–372 (1990)
4. Bowen, K.A., Kowalski, R.A.: Amalgamating language and metalanguage in logic programming. In: Clark, K., Tarnlund, S.A. (eds.) Logic Programming, Apic Studies in Data Processing, Academic Press, Inc, London (1983)
5. Brogi, A., Mancarella, P., Pedreschi, D., Turini, F.: Modular logic programming. ACM Trans. Program. Lang. Syst. 16(4), 1361–1398 (1994)
6. Codish, M., Debray, S.K., Giacobazzi, R.: Compositional analysis of modular logic programs. In: Proc. ACM Symp. on Principles of Programming Languages (POPL), pp. 451–464. ACM Press, New York (1993)
7. Giurca, A., Savulea, D.: An algebra of logic programs with applications in distributed environments. In: Annales of Craiova University. Mathematics and Computer Science Series, vol. XXVIII, pp. 147–159 (2001)
8. Henriksson, J., Aßmann, U., Heidenreich, F., Johannes, J., Zschaler, S.: How dark should a component black box be? The Reuseware Answer. In: ECOOP 2007. Proc. of the 12th International Workshop on Component-Oriented Programming (WCOP) co-located with 21st European Conf. on Object-Oriented Programming (to appear, 2007)
9. Henriksson, J., Johannes, J., Zschaler, S., Aßmann, U.: Reuseware – adding modularity to your language of choice. In: Proc. of TOOLS EUROPE 2007: Special Issue of the Journal of Object Technology (to appear, 2007)
10. Karali, I., Pelecanos, E., Halatsis, C.: A versatile module system for prolog mapped to flat prolog. In: Proc. ACM Symp. on Applied Computing (SAC), pp. 578–585. ACM Press, New York (1993)
11. Kifer, M., de Bruijn, J., Boley, H., Fensel, D.: A realistic architecture for the semantic web. In: RuleML, pp. 17–29 (2005)
12. Miller, D.: A theory of modules for logic programming. In: Proc. IEEE Symp. on Logic Programming, pp. 106–114. IEEE Computer Society Press, Los Alamitos (1986)
13. Sannella, D.T., Wallen, L.A.: A calculus for the construction of modular prolog programs. Journal of Logic Programming 12(1-2), 147–177 (1992)
14. Wirsing, M.: Structured algebraic specifications: A kernel language. Theoretical Computer Science 42(2), 123–244 (1986)

Towards Semantically Grounded Decision Rules Using ORM⁺

Yan Tang, Peter Spyns, and Robert Meersman

Semantics Technology and Applications Research Laboratory
Department of Computer Science
Vrije Universiteit Brussel
Pleinlaan 2, B-1050 BRUSSEL 5, Belgium
{yan.tang,peter.spyns,robert.meersman}@vub.ac.be

Abstract. Recently, ontologies are proposed for many purposes to assist decision making, such as representing terminology and categorizing information. Current ontology-based decision support systems mainly contain semantically rich decision rules. In order to ground the semantics, we formalize those rules by committing them to domain ontologies. Those semantically grounded decision rules can represent the semantics precisely, thus improve the functionalities of many available rule engines. We model and visualize the rules by means of a novel extension of ORM. These rules are further stored in an XML-based markup language, ORM⁺ ML, which is a hybrid language of Rule-ML and ORM-ML. We demonstrate in the field of on-line customer management.

Keywords: semantics, ontologies, DOGMA, ORM, markup language.

1 Introduction

Nowadays, ontologies [7, 6] are used to support collaborative decision making systems, such as [22]. These ontologies contain a domain vocabulary where the concepts provide semantically precise domain knowledge.

In [8], Guarino and Poli distinguish the *ontological level* and other levels, such as the *logical level*, in Knowledge Representation (KR) formalisms according to the kinds of primitives used. The *ontological level* is the level of *meaning*, which is accounted for by a set of *axioms* and a vocabulary. Mizoguchi and Ikeda further argue that an *axiom* in a formal ontology is defined as declaratively and rigorously represented knowledge, which is agreed and (*must* be) accepted in a *world* (or a *context*) [16]. For example, human beings *must* have *red* blood. This fact is formally agreed by the domain experts in the biological domain. The ontology-based applications (or tasks) in a domain use this kind of *axiomatized facts*. We argue that the applications can *override* but *not change* the axiomatized facts of the domain ontology. For an *axiom* is generally accepted to be true. In addition, we also argue that the applications and tasks in a domain may influence the *evolution* of the ontology. The principle of modeling the ontologies with respect to the scope of the

A. Paschke and Y. Biletskiy (Eds.): RuleML 2007, LNCS 4824, pp. 78–91, 2007.

applications and tasks can be found in several literatures, such as [5, 20, 23]. In this paper, we assume the notion of 'domain ontology' is *ideal*, which means that the ontology defines *all* the concepts used by *all* the applications in a domain.

At the *logical* level, primitives are propositions, predicates, logical functions and operators, which describe the relations[1] among the objects in the real world. For example, a 'bottle' can hold 'water', 'wine' or something else. According to Guarino et. al. [8], the logical level is considered as a separated level from the ontological level. It is because "... no particular assumption is made however on the nature of such relations, which are completely general and content-independent... the logical level is the level of formalization: it allows for a formal interpretation of the primitives, but their interpretation is however totally arbitrary..." Therefore, we shall consider the *ontological axioms* separately from the logical *functional rules* and the *logical operators* (the "application rules"). The *logical relations* and *operators* that construct the relations between those concepts (or "objects") should be placed between an ontology base level and the application level.

On the one hand, the *ontological level* is separated from the *logical level*. But on the other hand, to establish a link between these two levels is important. Some logical operators, such as those used to connect high level rules in [2], are essential to capture *formal semantics* during the rule modeling activities. A proper link between the ontological level and the logical level needs to be established. One way of linking these two layers is to lay the logical level above the ontological level as Berners-Lee et al. did in [1]. In this paper, we refine the logical level as a part of *commitment layer*. The commitment layer, which will be discussed in this paper, resides between the levels of ontology base and applications. It specifies the *semantic interpretation* of the ontology base for the applications, such as the decision making.

During the past EU project FF-POIROT[2], a strong emphasis was placed on how to model and store the commitments for the decision rules in a sharable way. In the earlier papers, Zhao et al. [23, 13] addresses the question of how to model and use ontology-based rules. The issues of formally committing, storing and visualizing them were not yet covered. Based on his work, we try to tackle the problems of how to *commit, visualize* and *store* semantically grounded decision rules in this paper.

The approach studied in this paper is not restricted to the decision making domain. Instead, we consider it as a use case to introduce our approach to the commitment.

We organize the paper as follows: In section 2, we explain the background of ontology and its commitment for the decision rules. ORM (Object Role Modeling, [9]) diagram and ORM ML (markup language, [4]) are used to model and store these commitments. In section 3, we try to extend ORM for the visualization, the result of which is called ORM⁺. We design a markup language for ORM⁺ (ORM⁺ ML) in order to store those rules in a *sharable* way. A use case in the field of Human Resource Management is illustrated. We present related work and open a discussion in section 4. We summarize the paper and illustrate our future work in section 5.

---

[1] We argue that the relations discussed here are not *formal ontological relations*. Guarino describes formal ontological roles in *mereology* (the theory of the part-whole relation), *topology* (intended as the theory of the connection relation) and so forth [7].

[2] The FF-POIROT (Financial Fraud Prevention-Oriented Information Resources using Ontology Technology) project aimed to represent a relevant body of (financially related) judicial knowledge in ontologies. http://www.ffpoirot.org/

2 Background: ORM Approach to Commitment

An ontology is a semiotic representation of agreed conceptualization in a subject domain [6, 7]. As an ontology representation framework and ontology engineering methodology, DOGMA[3] (Developing Ontology-Grounded Methods and Applications [15, 19]) was designed as a methodological framework inspired by the tried-and-tested principles from conceptual database modeling. We transport the principle of data independence (as applied in modern database) into the principle of meaning independence (to be applied for ontologies). An application *commits* its local vocabulary to the meaning of the ontology vocabulary. In the DOGMA framework one constructs (or converts) ontologies by *double articulation*: the ontology base layer that contains a vocabulary of simple facts called *lexons*, and the *commitment* layer that formally defines rules and constraints by which an application (or "agent") may make use of these lexons.

A lexon is a quintuple $< \gamma, t_1, r_1, r_2, t_2 >$, where t_1 and t_2 are the terms that represent two concepts in some language, e.g. English. r_1, r_2 are the roles referring to the relationships that the concepts share with respect to one another. γ is a context identifier which is assumed to point to a resource, and which serves to disambiguate the terms t_1, t_2 into the intended concepts, and in which the roles r_1, r_2 become "meaningful". For example, a lexon *<Order Manager, accept, is Accepted by, Customer Request>*[4] explains a fact that "(an) order manager accepts (a) customer request".

A *commitment*, which corresponds to an explicit instance of an intentional logical theory interpretation of the applications, contains a set of rules in a given syntax and describes a particular application view of reality, such as the use by the application of the (meta-) lexons in the ontology base. The commitment layer provides the possibility of having *multiple* views on and uses of the same stored lexons. By modeling the commitments, the "meaning space" can be subdivided in different ways according to the *need*s of a specific application that will add its particular *semantic restrictions*. This process is also called '*to commit ontologically*'.

The commitments need to be expressed in a *commitment language* that can be interpreted. A commitment language in [3] initially explicates commitment from the perspective of *semantic path*, which can be restricted by relational database constraints.

The semantic path provides the *construction* direction of a lexon. Suppose we have a lexon <Order Manager, accept, is Accepted By, Customer Request>, which has the constraints as "*one* customer request is accepted by *at most one* order manager". We apply the uniqueness constraints *UNIQ* on the path written as below:

```
p1 = [Customer Request, is Accepted By, accept, Order
Manager]: UNIQ (p1).
```

---

[3] We emphasize that the DOGMA is not 'dogmatic'.

[4] In this paper, we do not focus on the discussion of the context identifier γ. We omit it in the actual lexons. E.g. *<γ, Order Manager, accept, is Accepted by, Customer Request>* is written as *< Order Manager, accept, is Accepted by, Customer Request >*.

The uniqueness constraints are borrowed from the database uniqueness constraints. In order to 'commit ontologically', those constraints (e.g. UNIQ) are used when a decision rule need to be committed to an ontology.

ORM (Object Role Modeling, [9]) was originally intended for modeling and querying databases at the conceptual level, enabling modeling, validating and maintaining database processes. In [19, 4], the authors studied many advantages of ORM as a semantically rich modeling manner. Furthermore, they argue that it is rather feasible to model and visualize the commitments using ORM. For ORM has an expressive capability in its graphical notation and verbalization possibilities. It enables non-technical domain experts to communicate each other with ORM diagrams (E.g. Fig. 1). Later, Demey et al. present an XML-based ORM markup language (ORM-ML, [4]), which enables exchanging ORM models including ORM-syntax application rules. The ORM-ML can be further mapped into OWL[5], which makes it possible to adapt many available ontology technologies.

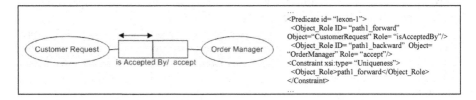

Fig. 1. An ORM diagram example[6] (left) and its ORM-ML[7] representation (right): *each* customer request is accepted by *at most one* order manager

However, ORM still lacks several logical operators and connectors while grounding the semantics for the decision rules, e.g. the Sequence and the Implication operators. In the following sections, we will focus on extending ORM and ORM-ML for visualizing, committing and storing the decision rules.

3   ORM+ Approach to Commitment

We adopt the basic propositional logic connectives in [17], the modality connectives in [12] and introduce a new operator – the *sequence* operator for ORM+. We illustrate the graphical presentation together with the ORM+ Markup Language – ORM+ ML. The ORM+ ML is a standard language for storing decision rules in extended ORM (ORM+) models on the basis of combining Rule-ML[8] and ORM-ML [4].

In [17], Nissanke lists 5 basic propositional logic connectives: *negation, conjunction, disjunction, implication* and *equivalence*. As the equivalence connective is often used for checking the equivalence of two logical propositions or to prove a

---

[5] OWL stands for Web Ontology Language. http://www.w3.org/TR/owl-features/
[6] We do not use ORM2 notations here. In ORM2, the uniqueness constraint is modeled with a bar without two arrows.
[7] All the XML files illustrated in this paper can be download at:
http://www.starlab.vub.ac.be/website/ORMplus/ml/examples
[8] http://www.ruleml.org/

logical theory but not for modeling the commitments, we exclude it here. In what follows, we will discuss the usage of other four basic propositional logic connectives.

1. Negation: If we want to add the negation operator[9] to a fact, e.g. "the customer request is *not* accepted by order manager", we apply the *negation* constraint to a commitment as:

```
p1 = [Customer Request, is Accepted By, accept, Order
Manager]: NOT[10] (p1(is Accepted By), p1(accept)).
```

The 'NOT' operator is applied to lexon (co-)roles. A lexon (co-)role is specified by a path label that is followed by a role in the parentheses, e.g. p1(is Accepted By).

The negation connective is not a constraint in ORM. However, it is possible to model negation propositions. In ORM, one uses "closed-world" and specific "open-world" assumptions (page 61, [9]). The "closed-world" assumption uses the *absence* of positive information (e.g. Customer request is accepted by Order manager) to imply the negative (e.g. Customer request is not accepted by Order manager). With an "open-world" approach, negative information is explicitly stored using negative predicates or status object types. I.e. "Customer request" has Acceptance status {'Accepted', 'Not Accepted'} and *each* "Customer request" has *at most one* Acceptance status (Fig. 2, A). Or, "Customer request" has two subtypes "Accepted customer request" and "Unaccepted customer request", which are mutually exclusive (Fig. 2, B).

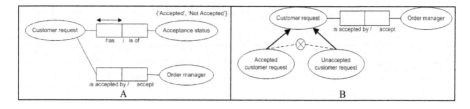

Fig. 2. Modeling negation in ORM

ORM uses curly braces to list all possible values of a value type. The schema in Fig. 2 (A) can be explained as: *each* "Customer request" has *at most one* "Acceptance status", which has the value "Accepted" or "Not Accepted"; "Customer request" is accepted by "Order manager". ORM uses a circled X symbol "⊗" to indicate the *exclusion constraint*. One direction arrow-tipped bar is used to model the *subtype* constraint. The schema design in Fig. 2 (B) can be verbalized as: "Customer request" is accepted by "Order manager"; "Customer request" has two subtypes - "Accepted

---

[9] In this paper, the connectives and operators are applied at the decision item level, not at the decision rule level. A decision rule is constructed with a set of decision items. We separate a decision item from a decision rule because a decision rule is non-monotonic but a decision item can be monotonic.

[10] The syntax and semantic of the commitment language used in this paper can be found at: http://starlab.vub.ac.be/website/SDT.commitment.example

customer request" and "Unaccepted customer request". These two subtypes are mutually exclusive.

Although transferring the negation connective to the value constraint or the exclusive constraint doesn't lose any information of a negative proposition in ORM, it is still not easy for a domain expert to know when the negative status is taken. For both positive and negative statuses of a type are modeled in the same schema. The domain expert has to make the analysis or reckon on extra information in order to know whether he uses the negative status or the positive one. To simplify the situation, we introduce the negation constraint in the ORM⁺.

The example commitment is visualized and stored in an XML file (Fig. 3). We visualize a *negation* operator by marking a "¬" to the applied role(s), which are enclosed in a dotted rectangle. In the resulting ORM⁺ ML file, we use the tag pair *<Neg/>* to store the negation operator. The element 'Neg' is borrowed from FOL RuleML V0.9[11] in order to store the classical negation operator.

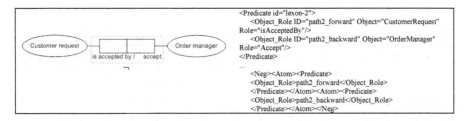

Fig. 3. A customer request is *NOT* accept by an order manager. An order manger does NOT accept a customer request.

2. Conjunction: The *conjunction* binary operator is to construct a logical *AND* with the conjunction operator ∧. For example, the decision item "customer is listed in a customer catalog *AND* customer state is normal". We illustrated the commitment as:

```
(p2=[Customer, is Listed In, list, Customer Catalog],
 p3=[Customer, has, is Of, Normal State]): AND(p2, p3).
```

In this example, the conjunction constraint AND is applied to the semantic path level. The conjunction operator is equivalently used as *set intersection* in the set theory. A traditional way of expressing set intersection in ORM is to use set comparison constrains (such as subset, equality and exclusion constraints) to restrict the way of the population of one role[12], or role sequence (page 229, [9]). Set intersection action is executed at the *query* level. That is to say, ORM itself doesn't provide modeling techniques for the conjunction operator.

---

[11] The first order logic web language: http://www.ruleml.org/fol/

[12] Note that the DOGMA (co-)role is defined differently from ORM role. In ORM, a type (object type or relationship type) is a set of all possible instances. For a given database schema, types are fixed. A fact type has several roles; each role is associated with a column in its fact table. In DOGMA, (co-)role is used to describe a formal ontological relation for a binary fact. E.g. a role "is-a" is a taxonomical relation that describes the ontological subsumption relationship.

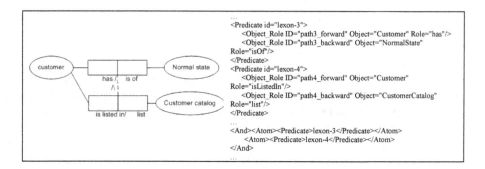

Fig. 4. A customer is listed in a customer catalog *AND* his customer state is normal

In the settings of our use case, the commitments mainly use the conjunction connective to deal with the deductive proofs for the decision rules. For example, it is used to eliminate rule duplications or violation. Therefore, we need to specifically introduce a generic conjunction operator in ORM⁺ for the formal commitments.

Graphically, the dotted bar marked with a "∧" denote the *conjunction* operator (Fig. 4). In the XML file, it is stored between the tag pair *</And>*, which is borrowed from FOL (First Order Logic) RuleML V0.9.

3. Disjunction: The *disjunction* operator is to link multiple choices of several lexons with the logical disjunction operator ∨. Such rules are often indicated with "OR" in a natural language. I.e. we have a commitment as:

```
(p4 = [Customer Request, is Refused By, refuse, Order
Manager], p5 = [Customer Request, is Canceled By,
Cancel, Order Manager]): OR (p4, p5).
```

The commitment describes a decision item - "customer request is refused OR canceled by order manger". Similar to the conjunction constraint, the disjunction constraint OR is also applied to the path level. In the field of Boolean algebra and decision rules, the choice of either of these is called *exclusive or*. The operator of the intersection of these is called *inclusive or*, which has been discussed as the *conjunction* operator in the previous subsection. In this section, we use the disjunction operator to formulate the *inclusive or* constraint.

In ORM, the *exclusive constraint* is used to indicate that two roles are mutually exclusive. In Fig. 5, the (co-)role pairs "is refused by/ refuse" and "is canceled

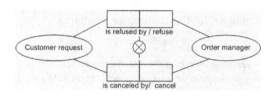

Fig. 5. Modeling disjunction in ORM

```
    ...
<Predicate id="lexon-5">
    <Object_Role ID="path5_forward" Object="CustomerRequest" Role="isRefusedBy"/>
    <Object_Role ID="path5_backward" Object="OrderManager" Role="refuse"/>
</Predicate>
<Predicate id="lexon-6">
    <Object_Role ID="path6_forward" Object="CustomerRequest" Role="isCanceledBy"/>
    <Object_Role ID="path6_backward" Object="OrderManager" Role="cancel"/>
</Predicate>
    ...
<Or><Atom><Predicate>lexon-5</Predicate></Atom>
    <Atom><Predicate>lexon-6</Predicate></Atom></Or>
    ...
```

Fig. 6. A customer request is refused OR canceled by an order manger

by/cancel" are mutually exclusive. The schema is verbalized as: *no* "Customer request" is refused *and also* is canceled by *an* "Order manager".

The ORM⁺ diagram reuses the exclusive constraint from ORM to model the disjunction operator. In the ORM⁺ ML file, we store the exclusive constraint between the tag pair *</Or>*, which is taken from FOL RuleML V0.9 (Fig. 6).

4. Implication: We use the *implication* operator to construct one decision for a decision rule, which in natural language contains the keywords "IF...THEN... ELSE". For example, the decision "*IF* a customer is not listed in the customer catalog *THEN* order manager creates new customer" can be expressed as a commitment:

```
(p6 = [Customer, is Listed In, list, Customer Catalog],
p7 = [Order Manager, create, is Created By, New
Customer]): IMP(NOT(p6(is Listed In),p6(list)),p7).
```

In the commitment, we use IMP to indicate the implication connectives for two (constrained) semantic paths. In the ORM⁺ diagram, we use the symbol $\Rightarrow$ to designate the *implication* logical operator (Fig. 7). In the ORM⁺ XML file, a rule tag pair with the attribute value *<Rule xsi:type = "Implies">* is used.

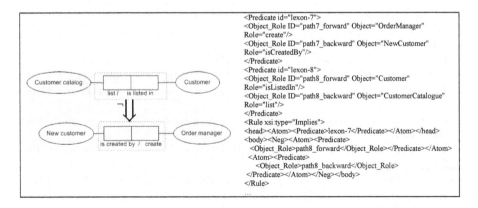

Fig. 7. *IF* a customer is *NOT* listed in a customer catalog (a customer catalog does NOT list a customer), *THEN* an order manager creates a new customer

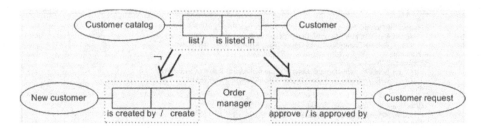

Fig. 8. *IF* a customer is NOT listed in a customer catalog, *THEN* an order manager creates a new customer, *ELSE* the order manager approves the customer request

```
...
<Predicate id="lexon-7">
<Object_Role ID="path7_forward" Object="OrderManager" Role="create"/>
<Object_Role ID="path7_backward" Object="NewCustomer" Role="isCreatedBy"/>
</Predicate>
<Predicate id="lexon-8">
<Object_Role ID="path8_forward" Object="Customer" Role="isListedIn"/>
<Object_Role ID="path8_backward" Object="CustomerCatalogue" Role="list"/>
</Predicate>
<Predicate id="lexon-9">
<Object_Role ID="path9_forward" Object="OrderManager" Role="approve"/>
<Object_Role ID="path9_backward" Object="CustomerRequest" Role="isApprovedBy"/>
</Predicate>
<Rule type="Implies">
   <head>
     <Atom><Predicate>lexon-7</Predicate></Atom>
   </head>
   <body> <Neg>
         <Atom><Predicate><Object_Role>path8-forward</Object_Role></Predicate></Atom>
         <Atom><Predicate><Object_Role>path8-backward</Object_Role></Predicate></Atom>
   </Neg></body>
</Rule>
<Rule type="Implies">
   <head><Atom><Predicate>lexon-9</Predicate></Atom></head>
   <body><Atom><Predicate>lexon-8</Predicate></Atom></body>
</Rule>...
```

Fig. 9. The ORM-ML file for Fig. 8

Fig. 7 shows a branch of decision item. In practice, a traditional decision rule includes at least two decision branches as it is non-monotonic. Fig. 8 and Fig. 9 are such an example.

In the following two sections, we will illustrate two important operators in the Modal Logic – the Necessity and Possibility. The semantics of modal logics has been studied so far by logicians and physiologists, such as the possible world semantics by Saul Kripke [14].

In the research field of object-role molding, Halpin categorizes rule modalities into *alethic* and *deontic* [12]. Alethic rules "impose necessities, which cannot be violated by any applications because of physical or logical law...", e.g. there is only one sun in the sky. A deontic rule "imposes obligations, which may be violated, even though they ought not ...", e.g. no one is allowed to kill another person. In this paper, we focus on the alethic modal operators discussed by Halpin: the *necessity* operator ϒ and the *possibility* operator ◊.

5. Necessity: In ORM 2 [11], an alethic modality of *necessity* ϒ is used for positive verbalizations by default. For example, the fact 'a customer is listed in a customer catalog' may be explicitly verbalized as 'a customer is NECESSARILY listed in a customer catalog' by default. Halpin interprets it in terms of *possible world semantics*, which are introduced by Saul Kripke et al. in the 50's. A proposition is *"necessarily* true *if and only if* it is true in all possible worlds". The *facts* and *static constraints* belong to a possible world, in which they *must* exist at some point in time. Therefore, the necessity operator may explicitly append on the fact 'a customer is listed in a customer catalog' by default.

Following the same example, we get the commitment as:

```
p8 = [Customer, is Listed In, list, Customer Catalog]
:L(p8(is Listed In), p8(list)).
```

In this commitment example, 'L' is used to constrain the lexon (co-)role for a semantic path. We visualize a *necessity* operator in ORM⁺ by marking a "Υ" to the applied role(s), e.g. "is listed by" and "list" shown in Fig. 10. In the resulting ORM⁺ ML file, we use the tag pair *<Constraint/>* with the attribute *xsi:type= "Necessity"* to store this commitment.

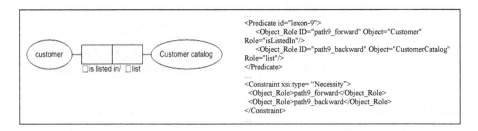

Fig. 10. A customer is NECESSARILY listed in a customer catalog. A customer catalog NECESSARILY list(s) a customer.

6. Possibility: The alethic modality of possibility ◊ is often verbalized as 'possibly'. Formally speaking, a proposition is *possible* "if and only if it is true in *at least one possible* world; impossible propositions are true in no possible world" [12].

For instance, we have a rule 'a customer POSSIBLY has a normal state', which is expressed as a commitment:

```
p9 = [Customer, has, is Of, Normal State]
:M(p9(has), p9(is Of)).
```

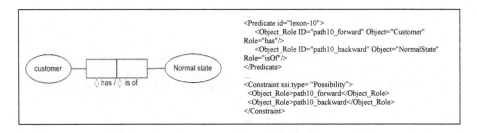

Fig. 11. A customer POSSIBLY has a normal state. A normal state POSSIBLY is of a customer.

As the necessity operator, the possibility operator 'M' in the commitment is used to constrain a lexon (co-)role for a semantic path. It is visualized in ORM⁺ by marking ◊ near the applied roles, "has" and "is of" (Fig. 11). In the resulting ORM⁺ ML, the tag *<Constraint/>* with the attribute value *xsi:type= "Possibility"* are used to store the commitment of possibilities.

7. Sequence: The *sequence* operator indicates execution (time) sequence of two rule units. It is mentioned in our previous work in [23]. We intend to use the sequence operator to reason on the execution of processes. For example, we have a rule "an order manager verifies a customer request AFTER the order manager receives the customer request", which constraints the time sequence of two processes. The commitment is formalized as the following:

```
(p10 = [Order Manager, verify, is Verified By, Customer
Request], p11 = [Order Manager, receive, is Received
By, Customer Request]): ORDER(p10, p11).
```

In the commitment, 'ORDER' is used to indicate the sequence of several semantic paths. A dotted arrow marked with ">>" is used to visualize the *sequence* operator. We place an event that happens before another one at the starting point of this arrow. The following event is placed at the ending point of the arrow (Fig. 12). In its XML file, the *sequence* rule is indicated in the attribute type of the tag *<Rule/>* together with the *direction* notification attribute, e.g. <Rule xsi:type="Sequence" direction="forward">.

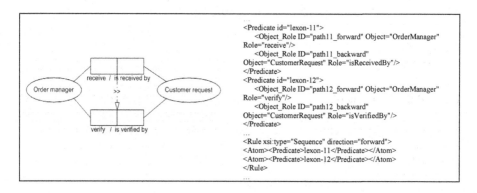

Fig. 12. An order manager verifies a customer request AFTER the order manager receives the customer request

In this section, we focus on how to visualize decision commitments with a novel extension of ORM approach, and how to store it in a XML file. We have developed a JAXB[13] (Java Architecture for XML Binding) package for the latest ORM⁺ ML Schema (V1.0). It is an open Java API[14] to create and read ORM⁺ XML files.

---

[13] http://java.sun.com/developer/technicalArticles/WebServices/jaxb/
[14] Download: http://www.starlab.vub.ac.be/website/files/ORMplus%20JAXB%20package.zip

4   Related Work and Discussion

Semantically grounded decision rules visualized in extended ORM can also be represented in UML(Unified Modeling Language[15]). UML can be used as an ontology (and commitment) modeling language [2] – thus to capture domain (e.g. business domain) semantics and ground them into ontologies is possible. However, UML has its own limitation as a commitment modeling language. In [10], Halpin discussed that UML provides very limited support of ontological *subset* and *exclusion* constraints.

The OMG (Object Management Group) has modeled the semantics for business vocabulary and rules (SBVR, [18]). They provide the definition of logical operators and their verbalization in natural language. And, ORM2 is used to visualize their business models (Annex I, [18]). One models the business rules, which contain the implication operator, by making an extra link to a verbalized sentence. As discussed in this paper (section 3), it is not a best solution to model such an operator properly.

The ORM⁺ Markup Language is designed on the basis of combining Rule-ML[16] and ORM-ML [4]. As Rule-ML and ORM-ML design their syntax and language semantics based on XML syntax, we design the *ORM⁺ ML* constraints and their attributes through a normative XML schema. In ORM-ML Version1.3[17], we have defined 23 ORM constraints. We add 4 logical operators (*And, Or, Neg, Implies*) from FOL (First-Order Logic Web Language) Version 0.9[18], 2 modality operators in [12] and the *Sequence* operator for *ORM⁺ ML*[19]. The *ORM⁺ ML* schema reuses 31 type definitions of *ORM ML* schema and 10 type definitions of *FOL Rule-ML* schema.

Commitments stored in ORM⁺ ML can be used by the inference rule engines. Many rule engines were developed for specific purposes, such as: 1) InfoSapient[20] that is to deploy and maintain operation rules by applying fuzzy logic; 2) JLisa[21], which is to access java programming code by providing rules in Clips (a productive development and delivery expert system tool[22]) framework and validate rules; 3) OpenRules[23] that provides methodology, services and tools for business (decision) rule maintenance and deployment; 4) OpenLexicon[24] that is an open-source online business decision rule editing. Our semantically grounded decision rules, which are committed properly to a domain ontology, represent semantics precisely and thus can improve the functionalities of these rule engines mentioned above. An item of our future work is to map the ORM⁺ ML to one of the listed rule engines.

Current commitments in DOGMA contain very rich constraints and axioms. Including the propositional logic connectives, modal logic operators and the sequence connective introduced in the paper, current commitments comprise all ORM

---

[15] http://www.omg.org/technology/documents/formal/uml.html
[16] http://www.ruleml.org/
[17] The ORM-ML Schema V1.3: http://www.starlab.vub.ac.be/mustafa/ormml.v1.3.xsd
[18] http://www.ruleml.org/fol/
[19] The ORM⁺ ML Schema can be downloaded at:
   http://www.starlab.vub.ac.be/website/ORMplus/ml
[20] http://info-sapient.sourceforge.net/
[21] http://jlisa.sourceforge.net/
[22] http://www.ghg.net/clips/WhatIsCLIPS.html
[23] http://openrules.com/
[24] http://openlexicon.org/blog/

constraints. However, not all kinds of decision rules can be formalized in the commitment, e.g. fuzzy rules. It is a limitation of ORM⁺.

5 Conclusion and Future Work

This paper addresses many challenges of managing *semantically grounded* decision rules in a large *community*: modeling, committing, visualization and sharing. We have focused on the topics of how to *commit* semantics of decision rules, how to *visualize* and to *store* them in a new format. The *commitment* is designed and formalized based on DOGMA (Developing Ontology-Grounded Methods and Applications) framework. We use ORM (Object Role Modeling) approach to support modeling commitment. In order to support differentiate decision semantics, we tried to extend ORM, the extension of which is called ORM⁺. We tried to store ORM⁺ diagrams in ORM⁺ ML to make them sharable among software agents.

The construction of ORM⁺ proposed in this paper is intended to *model, store* and *share* semantically grounded decision rules among a decision group. We're recently working on the theory and the implementation of Semantic Decision Table (SDT, [22]), which enriches traditional decision tables with explicit semantics. While constructing a SDT among a decision group, the approach in this paper can be adopted for grounding the semantics of decision rules.

We have developed a tool – DOGMA Studio Workbench[25] plug-in, which takes formalized commitments as the input and generates ORM⁺ files as the output. The future work is to develop a tool that supports visualization of ORM⁺. In our early work [21], we proposed to use Prolog language as a reasoning language for the commitments in ORM. In the future, we will design a tool, which translates the ORM⁺ commitments into Prolog code to reason them. By considering the SDT as a 'mini' rule-based decision making system, we will also develop an open-source java package for the mapping between ORM-ML and SDT reasoner.

Acknowledgments. We hereby express our appreciation for Damien Trog for reviewing the paper. The research is partly funded by the EU FP6 PROLIX project.

References

1. Berners-Lee, T., Hendler, J., Lassila, O.: The Semantic Web, Scientific America (2001), URL: http://www.sciam.com/article.cfm?articleID=00048144-10D2-1C70-84A9809EC-588EF21&catID=2
2. Cranefield, S., Purvis, M.: UML as an ontology modelling language. In: IJCAI 1999. Proceedings of the Workshop on Intelligent Information Integration, 16th International Joint Conference on Artificial Intelligence (1999)
3. De Leenheer, P., de Moor, A., Meersman, R.: Context Dependency Management in Ontology Engineering: a Formal Approach. In: Spaccapietra, S., Atzeni, P., Fages, F., Hacid, M.-S., Kifer, M., Mylopoulos, J., Pernici, B., Shvaiko, P., Trujillo, J., Zaihrayeu, I. (eds.) Journal on Data Semantics VIII. LNCS, vol. 4380, pp. 26–56. Springer, Heidelberg (2007)

---

[25] The tool can be downloaded when the DOGMA studio workbench is installed.

4. Demey, J., Jarrar, M., Meersman, R.: Markup Language for ORM Business Rules. In: ISWC 2002. proc. Of International Workshop on Rule Markup Languages for Business Rules on the Semantic Web (2002)
5. Díaz-Agudo, B., González-Calero, A.: CBROnto: A Task/Method Ontology for CBR. In: Proceedings of the Fifteenth International Florida Artificial Intelligence Research Society Conference, pp. 101–105. AAAI Press, Stanford, California, USA (2002)
6. Gruber, T.R.: A translation approach to portable ontologies. Knowledge Acquisition 5(2), 199–220 (1993)
7. Guarino, N.: Formal Ontology and Information Systems. In: Proceedings of FOIS 1998, pp. 3–15. IOS press, Amsterdam (1998)
8. Guarino, N., Poli, R. (eds.): Formal Ontology in Conceptual Analysis and Knowledge Representation (Special issue of the International Journal of Human and Computer Studies), vol. 43(5/6). Academic Press, London (1995)
9. Halpin, T.A.: Information Modeling and Relational Databases: From Conceptual Analysis to Logical Design. Morgan Kaufman Publishers, San Francisco, California (2001)
10. Halpin, T.A.: Verbalizing Business Rules: Part 7. Business Rules Journal 5(7) (2004)
11. Halpin, T.A.: ORM 2. In: Meersman, R., Tari, Z., Herrero, P. (eds.) On the Move to Meaningful Internet Systems 2005: OTM 2005 Workshops. LNCS, vol. 3762, pp. 676–687. Springer, Heidelberg (2005)
12. Halpin, T.A.: Business Rule Modality. In: EMMSAD 2006. Proc. Of Eleventh Workshop on Exploring Modeling Methods for Systems Analysis and Design (2006), http://www.orm.net/pdf/RuleModality.pdf
13. Kerremans, K., Tang, Y., Temmerman, R., Zhao, G.: Towards ontology-based email fraud detection. In: Bento, C., Cardoso, A., Dias, G. (eds.) EPIA 2005. LNCS (LNAI), vol. 3808, Springer, Heidelberg (2005)
14. Kripke, S.: Semantical Considerations on Modal Logic. APF 16, 83–94 (1963)
15. Meersman, R.: Ontologies and Databases: More than a Fleeting Resemblance. OES/SEO 2001 Rome Workshop. Luiss Publications (2001)
16. Mizoguchi, R., Ikeda, M.: Towards Ontology Engineering. In: Proc. of The Joint 1997 Pacific Asian Conference on Expert systems / Singapore International Conference on Intelligent Systems, pp. 259-266 (1997)
17. Nissanke, N.: Introductory Logic and Sets for Computer Scientists. Addison Wesley Longman Ltd, Redwood City,CA, USA (1999)
18. OMG (Object Management Group), Semantics of Business Vocabulary and Business Rules Specification (2006), http://www.omg.org/docs/dtc/06-03-02.pdf
19. Spyns, P., Meersman, R., Jarrar, M.: Data modelling versus Ontology engineering. SIGMOD Record: Special Issue on Semantic Web and Data Management 31(4), 12–17 (2002)
20. Sure, Y., Staab, S., Studer, R.: Methodology for Development and Employment of Ontology Based Knowledge Management Applications. SIGMOD Record 31(4), 18–23 (2002)
21. Tang, Y., Meersman, R.: Towards building semantic decision table with domain ontologies. In: Man-chung, C., Liu, J.N.K., Cheung, R., Zhou, J. (eds.) ICITM 2007. Proceedings of International Conference of information Technology and Management, pp. 14–21. ISM Press (2007) ISBN 988-97311-5-0
22. Tang, Y., Meersman, R.: On constructing semantic decision tables. In: Wagner, R., Revell, N., Pernul, G. (eds.) DEXA 2007, Regensburg, Germany, vol. 4653, pp. 34–44. Springer, Heidelberg (2007)
23. Zhao, G., Gao, Y., Meersman, R.: An Ontology-based approach to business modelling. In: ICKEDS 2004. Proceedings of the International Conference of Knowledge Engineering and Decision Support, pp. 213–221 (2004)

Towards Ontological Commitments with Ω-RIDL Markup Language

Damien Trog, Yan Tang, and Robert Meersman

Semantics Technology and Applications Laboratory (STARLab)
Department of Computer Science
Vrije Universiteit Brussel
Pleinlaan 2, B-1050 BRUSSELS 5, Belgium
{dtrog,ytang,meersman}@vub.ac.be

Abstract. In the DOGMA (Developing Ontology-Grounded Methods and Applications) ontology engineering approach, ontology construction starts from an uninterpreted base of elementary fact types, called lexons, which are mined from linguistic descriptions. Applications that ontologically commit to such a lexon base are assigned a formal semantics by mapping the application symbols to paths in this lexon base. Besides specifying which concepts are used, we restrict how they may be used and queried with semantic constraints, or rules, based on the fact-based database modeling method NIAM/ORM. Such ontological commitments are specified in the Ω-RIDL[1] language. In this paper we present the Ω-RIDL Markup Language and illustrate with a case from the field of Human Resources Management.

1 Introduction

The emergence of ontology-based application, such as the Semantic Web [1], has marked the importance of using domain ontologies in applications. These ontology-based applications are an extension of traditional ones, within which the information (or rather 'concepts') is explicitly defined. Many other ontology-based applications interweave domain ontologies and applications [25,31]. An application needs to communicate with existing domain ontologies. This communication layer between the application and the ontology is also called *commitment layer* [24]. How to formally commit application (e.g. rules and symbols) to domain ontologies is still a complicated problem.

Guarino asserts that "specifications of the tasks that define the *semantic interpretation* of the domain knowledge" should not be considered as an ontology [11]. It should "concern more than one particular task". Further, Guarino stresses that "ontologies belong to the knowledge level and they may depend on particular points of view. We must observe however that it is exactly the degree of such dependence which determines the reusability and therefore the value of an ontology". He then gives the definition of ontology as "an ontology

---

[1] Ω-RIDL: Ontology Reference and IDea Language.

A. Paschke and Y. Biletskiy (Eds.): RuleML 2007, LNCS 4824, pp. 92–106, 2007.

is an explicit, partial account of a conceptualization ... it is a logical theory that constrains the intended models of a logical language". The ontological commitments are often task dependent. Based on Guarino's discussion, we argue that the ontological commitments should not be considered at the ontology level. Rather, the commitment layer resides at the application layer in this sense.

Inspired by *tried-and-tested* principles from conceptual database modeling, DOGMA[2] is designed as a methodological framework for ontology engineering.

In DOGMA an ontological commitment serves several purposes:

- Selection from the base of elementary binary facts (also called lexons).
- Axiomatization of lexons by applying semantic constraints.
- Interpretation of lexons by mapping application symbols.
- Conceptual querying of heterogeneous data sources.

In this paper, we focus on the illustration of these purposes by introducing the commitment language Ω-RIDL and its markup language format Ω-RIDL ML[3]. We construct the paper as follows: in Sect. 2, we give the background of DOGMA and the RIDL language. In Sect. 3 the syntax and semantics of Ω-RIDL ML is explained. In Sect. 4 we discuss the advantages and the limitations of Ω-RIDL and compare our work with others. We end the paper with conclusions and future work in Sect. 5.

2   Background

In this section, we introduce the DOGMA approach to ontology engineering followed by a discussion of the RIDL language.

2.1   DOGMA Approach to Ontology Engineering

A classical definition of ontology, which is given by Gruber, describes an ontology as an explicit specification of a conceptualization [9]. Conceptualization is defined by Guarino as the intended models, within which a set of logical axioms are designed to account for the intended meaning of a vocabulary [12,10]. Within the context of information systems and knowledge management, an ontology is considered as a "particular knowledge base, describing facts assumed to be always true by a community (of domain experts), in virtue of the agreed-upon meaning of the vocabulary used" [12]. Hence, ontologies are actually knowledge based resources that represent agreed domain semantics.

In the context of information systems, the fact that one often considers databases as a kind of knowledge based resources leads to reuse database engineering principles for ontology engineering[4]. For example, OBSERVER is a prototype

---

[2] Acronym for Developing Ontology-Grounded Methods and Applications [20,24].

[3] The latest version of the Ω-RIDL schema can be downloaded from: `http://www.starlab.vub.ac.be/website/omega-ridl`

[4] For instance, we can reuse conceptual database modeling methods to model ontologies [24,16,3]. On the one hand, conceptual database modeling approach to ontology

system that supports querying of ontology repositories, regarding that an ontology is assembled comparably as a database [22].

DOGMA is an ontology approach and framework that is not restricted to a particular representation language. One of the most characteristic features of DOGMA is the use of a layered architecture, in particular the possibility of having multiple views on and uses of the same stored conceptualization. The *meaning space* can be subdivided in different ways according to the needs of a specific application that will add its particular semantic restrictions according to its intended usage [19]. This corresponds to Guarino's theoretical distinction between the conceptual and ontological levels. The ontological level is the level of commitments, i.e. restrictions on the semantics of the primitives while the definite cognitive interpretation of a primitive is situated on the conceptual level [13, p.633]. Hence, a DOGMA ontology consists of a *lexon base* layer and a *commitment layer*. This layering approach enhances the potential for re-use and allows for scalability in representing and reasoning about formal semantics [32]. In analogy with a principle stemming from the linguistics fields, this has been dubbed the *double articulation principle* [24]. This principle is an orthodox *model-theoretic* approach to ontology representation and development [24].

The *lexon base* holds (multiple) intuitive conceptualizations of a particular domain. Each conceptualization is simplified to a "representation-less" set of context-specific binary fact types called *lexons*. A lexon represents a plausible binary fact-type and is formally described as a 5-tuple $\langle \gamma, headterm, role, co-role, tailterm \rangle$, where γ is an abstract *context identifier*, lexically described by a string in some natural language, and is used to group lexons that are logically related to each other in the conceptualization of the domain. Intuitively, a lexon may be read as: within the context γ, the *head term* may have a relation with *tail term* in which it plays a *role*, and conversely, in which *tail term* plays a corresponding *co-role*. For example the lexon: $\langle$ *HRM rules, Actor, starring in, featuring, Movie* $\rangle$, can be read as: In the context *HRM rules, Actor* plays the role of *starring in* a *Movie* and the *Movie* plays the role of *featuring* an *Actor*.

The goal of the lexon base is to reach a common and agreed understanding about the ontology terminology and is thus aimed at human understanding. Natural language terms are associated, via the language and context combination, to a unique word sense represented by a concept label (e.g. the WordNet [8] identifier person#2). With each word sense, a gloss or explanatory information is associated that describes that notion. Notice that there is a m:n relationship between natural language terms and word senses (to account for synonymy, homonymy and translation equivalents) [23].

The *commitment layer*, with its formal constraints [15], is meant for interoperability issues between information systems, software agents and web services.

---

engineering is promising and feasible; but on the other hand, ontology modeling is NOT equivalent to database modeling. Unlike data models, the fundamental asset of ontologies is their relative independence of particular applications.

It consists of a finite set of axioms that specify which lexons of the lexon base are interpreted and how they are visible in the committing application, and (domain) rules that semantically constrain this interpretation. This separation also motivated because experience shows that it is much harder to reach an agreement on domain rules than on a conceptualization [21].

2.2   From RIDL to Ω-RIDL

RIDL [18,28] appeared first in the literature in 1980 and was used as a conceptual "interface" on top of a relational database. It was designed from scratch to become an integrated formal syntactic support for information and process analysis, semantic specification, constraint definition, and a query/update language at a conceptual level rather than at the logical "flat data" level.

The conceptual support for RIDL was provided by the *idea-bridge* model for conceptual schemas developed using the NIAM[5] [7,29] methodology, which later became ORM[6] [14].

The RIDL language was separated into RIDL\cns, the constraint definition part, and RIDL\qu, the query-and-update part. They were used by two (in general disjunctive) users: database engineers and end-users. Database engineers used RIDL\cns to formally express a conceptual data schema and its constraints. At compile time, such a conceptual schema was (semi-)automatically transformed into a relational database schema (normalized or not). The end-user used RIDL\qu, after the generated database was populated, to retrieve/update data at runtime through (possibly interactive) conceptual queries on the conceptual schema, instead of constructing SQL queries on the underlying relational database.

Although RIDL was intended for database modeling, we reuse most of the language for specifying ontological commitments. There are many parallels between ontology engineering and data modeling as both consist of conceptual relations and rules [20,16]. The main advantage of RIDL (and thus Ω-RIDL) is that it has a strong methodological grounding and it is very close to natural language. This allows for easier communication with domain experts to create and verify conceptualizations and domain rules. Besides this the NIAM methodology supports the generation of relational databases by a methodological and automated process [6], which provides many insights on how to commit relational database schemas to conceptual models (and thus ontologies).

Akin to its predecessor, Ω-RIDL constitutes two parts for possibly two distinct users. It is a language for *conceptual querying* of heterogeneous data sources by most probably agents roaming the web. On how these conceptual queries can be constructed we refer to earlier work by Verheyden et al [30]. Next, it is also a language for specifying ontological commitments which will be the task of an ontology engineer in agreement with domain experts. The latter is the focus of this paper.

---

[5] NIAM: Natural-language Information Analysis Method.
[6] ORM: Object Role Modeling.

3   Committing to Ontologies with Ω-RIDL

The commitment layer using Ω-RIDL mediates the ontology base and its applications. In the context of domain ontologies, an application system and its database schema can be assigned a formal semantics in the form of a first order interpretation. Initially, Ω-RIDL was introduced to capture this kind of semantics by (1) separating a mapping from the schema's symbols and relationships to a suitable ontology base expressed in lexical terms, and (2) expressions of how database constraints restrict the use of (or commit to) to the ontological concepts [24].

There are two version of Ω-RIDL, a pseudo natural language version and a machine processable XML version called Ω-RIDL ML (where ML stands for Markup Language). In this paper our focus is on explaining Ω-RIDL ML, however we always state the corresponding Ω-RIDL statements where possible.

The software suite supporting the DOGMA framework is called DOGMA Studio Workbench. In this suite ontological commitments can be created graphically using the T-Lex tool [27], which supports almost all ORM constraints in a tree visualization, called NORM trees. These graphical constraints, are translated to the corresponding Ω-RIDL. As Ω-RIDL is richer than ORM (which will become clear in the following sections), some constraints will still need to be added textually, e.g. constraints on paths longer than one lexon or nested subset constraints. The tool then outputs the commitment in the Ω-RIDL ML format, from where it can be further machine processed.

An ontological commitment defined in Ω-RIDL ML consists of five parts: (i) Selection: which lexons to commit to. (ii) Axiomatization: constrains the usage of the lexons. (iii) Interpretation: maps the application symbols to lexons. (iv) Articulation: gives meaning to concepts and their relationships[7]. (v) Conceptual Queries: stores queries of the application in terms of the ontology. The conceptual queries are stored in commitments to support reuse and sharing of queries.

Before entering the discussion of Ω-RIDL, we first describe a use case in the PoCeHRMOM[8] project. The use case is used to illustrate the syntax and semantics of Ω-RIDL and Ω-RIDL ML. Note that the use case is non-trivial, which we use to demonstrate how to design Ω-RIDL commitments in real cases.

3.1   Case Study: PoCeHRMOM

A requirement in the PoCeHRMOM project is to analyze the complexity of competencies for different jobs. The community involved in the project brings forward the following HRM (Human Resource Management) rules:

1. Common competency, such as linguistic skills and computer knowledge, is required in all kinds of expertise.

---

[7] Articulation (see Sect. 3.4) is not part of the commitment process, but because Ω-RIDL ML is an exchange format we want it to contain this information.

[8] The PoCeHRMOM project uses ontologies to enhance human resource management. URL: http://www.starlab.vub.ac.be/website/PoCehrMOM

2. Every competence has its own competence level, which is measured by a set of criteria. The criteria are given by O*NET[9] and must be assumed unmodifiable.

3. Competences are sometimes relevant to each other. Thus we need to establish the dependencies properly for the competences. For instance, if a person has a competency of speech clarity, then he has the competency of oral expression.

4. For each job type, all the competences are not equally important. Importance is defined by domain experts and/or end users. With regard to this requirement, we set up a Score of Importance for every competence.

   (a) For the case of predefining the Score of Importance, domain experts need to manually set it for each competence that describes a job. E.g. an actor should have the highest importance score (5) for oral expression.

   (b) For the case of defining the competence level by the end users, we allow the end users to define the importance score.

In the following sections, we specify the above HRM domain rules as ontological commitments in Ω-RIDL.

3.2   Selection of Lexons

The first thing we specify in an ontological commitment is the selection of lexons. This will be the *Universe of Discourse* (UoD). Note that we can select lexons from different contexts, but we will need to explicitly indicate how these may be combined. But first we explain how these lexons are selected and what a lexon base path is.

The selection of lexons are specified in the `lexonSet` element, which is a child of the `commitment` root element. A lexon is a sequence of five string types for context, head term, role, corole and tail term. The following is an example of a lexon in the lexon set:

```
<lexonSet>
  <lexon id="l1">
    <context>HRM rules</context>
    <headTerm>Competence</headTerm>
    <role>with</role> <coRole>of</coRole>
    <tailTerm>Competence Level</tailTerm>
  </lexon>
</lexonSet
```

Lexon Base Paths. Lexons do not have a direction, however when constructing a path it can be given two directions by either starting with the head term or tail term. We call this a *directed lexon* when used in a path. Semantics-expressing constraints will be applied on these paths.

A *lexon base path* (from hereafter referred to as a path) is a path that can be walked in the selection of lexons. It has the following form:

---

[9] O*NET is a free online resource describing general knowledge about job types, profiles and occupations. URL: http://online.onetcenter.org/

```
<xsd:element name="lexonBasePath">
  <xsd:complexType>
    <xsd:choice>
      <xsd:sequence maxOccurs="unbounded">
        <xsd:element ref="context"/>
        <xsd:element name="headTerm" type="xsd:string"/>
        <xsd:element name="role" type="xsd:string"/>
        <xsd:element name="coRole" type="xsd:string"/>
        <xsd:element name="tailTerm" type="xsd:string"/>
      </xsd:sequence>
      <xsd:sequence maxOccurs="1">
        <xsd:element ref="context"/>
        <xsd:element name="headTerm" type="xsd:string"/>
      </xsd:sequence>
    </xsd:choice>
  </xsd:complexType>
</xsd:element>
```

The shortest path is formed by a context and a term. Otherwise a path is one or more (concatenated) directed lexons. When paths are created by concatenating lexons, the following rules apply:

- Let l_n be the directed lexon that is at the tail of the path and l_{n+1} be the directed lexon that will be concatenated to this path. Also let $\langle \gamma_n, h_n, r_n, c_n, t_n \rangle$ be respectively the context, head term, role, co-role and tail term of the directed lexon l_n.
- If γ_n equals γ_{n+1}, then h_{n+1} must equal t_n or h_{n+1} must be a subtype of t_n[10].
- If γ_n does not equal γ_{n+1}, then the combinations $\langle \gamma_n, t_n \rangle$ and $\langle \gamma_{n+1}, h_{n+1} \rangle$ must have been either:
  - articulated to the same concept[11];
  - equated or subtyped in the context links section (discussed in the following paragraph).

Examples of these paths are shown in the following sections describing the semantic constraints.

Context Links. When lexons are selected from different contexts we need a way to allow inter-context combinations of these lexons. In the ideal case the terms of all the lexons are articulated to concepts with declared glosses as explained in Sect. 3.4. Then when context-term combinations refer to the same concept we know they are equivalent. However, this will often not be the case. When we want to indicate which terms are equivalent we specify this in the `contextLinkSet` element. Additionally we can also specify subtypes by the *is-a* relation. These *equals* and *is-a* relations are given a fixed interpretation, built-in[12] in the Ω-RIDL interpreter, and is used to check the validity of constructed

---

[10] Relation interpretations, such as the subtype relationship are explained in Sect. 3.5.

[11] Articulation of term-context pairs happens outside of the ontology proper and is discussed in Sect. 3.4.

[12] Assumed in this paper. The DOGMA formalism however allows such (sub-)interpreters to be also externally custom-defined as "plug-ins" for the Ω-RIDL interpreter.

paths. Context links are local to the commitment, as opposed to articulation of concepts which is global.

```
Context Link: Date in "Date and Time" equals Date in "HRM rules"
<contextLink relation="equals">
  <contextTermPairSource>
    <context>HRM rules</context>
    <term>Date</term>
  </contextTermPairSource>
  <contextTermPairTarget>
    <context>Date and Time</context>
    <term>Date</term>
  </contextTermPairTarget>
</contextLink>
```

3.3   Axiomatization: Semantic Constraints and Rules

The semantic constraints in Ω-RIDL account for the intended meaning of the given conceptualization by defining constraints on (the set of) its given paths. They are the rules intended by the application UoD, e.g. integrity constraints such as the uniqueness constraint. Those constraints are also called static constraints, which are applied to all the possible "populations" of the fact types. In this section, we group those static constraints into uniqueness constraints, mandatory constraints, value constraints and set constraints.

In Ω-RIDL ML constraints are stored in a `constraintSet` element. Constraints are applied to one or more `setExpression` elements (depending on the kind of constraint), which are expressions that return a set of elements. These can be paths, but can also be more complicated constructions by using nested set operators such as the union, intersection and minus operators. The only requirement is that a path is used as a descriptor for the constrained expression.

It is important to note that constraints specify only permitted states of affairs about application systems expressed in Ω-RIDL language. Whether these states hold or not depends on the actual implementations of such application systems. More specifically, the interpretation (evaluation if you like) of a path into a set of instances depends on the application system's instance at the moment of evaluation.

Uniqueness Constraints. The *uniqueness constraints* are often seen in natural language in the sentence structure of "each ... at most one ...". For example, "each competence has at most one competence level". We apply such a uniqueness constraint on a lexon ⟨HRML Rules, competence, with, of, competence level⟩ by writing the following Ω-RIDL statement and its equivalent in Ω-RIDL-ML:

```
in "HRML Rules" each Competence is with at most 1 "Competence Level"
<occurrenceConstraint comparator="lessThanOrEqual" number="1">
  <setExpression>
    <lexonBasePath>
      <context>HRM rules</context>
      <headTerm>Competence</headTerm>
      <role>with</role>  <coRole>of</coRole>
      <tailTerm>Competence Level</tailTerm>
    </lexonBasePath>
  </setExpression>
</occurrenceConstraint>
```

Observe that in the Ω-RIDL statement we added "is" before the "with" role name for better reading, but it is optional. We also omitted the co-role name "of". This is allowed when there is no ambiguity. For example if there are two relations between Competence and Competence Level where only the co-role differs, then the full lexon needs to be written: `each Competence is with / of at most one "Competence Level"`. Also note that term and role names comprising more than one word are surrounded by double quotes.

The uniqueness constraint is specified in the `occurrenceConstraint` element, which contains two attributes: `number` and `comparator`. In natural language, it indicates the occurrence of "competence level" can only be less than or equal to 1 when we say each competence has at most one competence level.

So far, the uniqueness constraint discussed is also called an internal (intrapredicate) uniqueness constraint ([14, p.129]). The internal uniqueness constraints are only applied on one path (in ORM only on one fact). There is another kind of uniqueness constraint: the external (interpredicate) uniqueness constraint, which is used to constrain a set of paths.

For example, for each date, the combination of the year, month and day may be unique in the given application context. In other words, the combination of year, month and day refers to only one date. The Ω-RIDL statement is written as:

```
in "Date and Time" each Date is identified by
(Year of Date and Month of Date and Day of Date)
<externalUniquenessConstraint>
  <setExpression>
    <lexonBasePath>
      <context>Date and Time</context>
      <headTerm>Year</headTerm>
      <role>of</role> <coRole>with</coRole>
      <tailTerm>Date</tailTerm>
    </lexonBasePath>
  </setExpression>
  <setExpression> ... </setExpression>
  <setExpression> ... </setExpression>
</externalUniquenessConstraint>
```

The combination of tail terms of the paths follow the same rules as those stated previously for path construction. In this example the tail terms of the paths is three times the same term, viz. Date.

In our examples of Ω-RIDL statements we have always prefixed our statements with the "in" construct, followed by the context in which the statement is executed. When there is only one context in the commitment this may be omitted, as in the following examples. In the Ω-RIDL ML we always explicitly state the context as it is machine generated.

Mandatory Constraints. A *mandatory constraint* is used when the total population of an object type plays a specific role. In natural language, we often verbalize the mandatory constraint in the sentence structure of "each ... at least one ...". For example, "each employee has at least one competency". In Ω-RIDL, we write:

```
each employee is with at least 1 competency
<occurrenceConstraint comparator="greaterThanOrEqual" number="1">
  ...
</occurrenceConstraint>
```

We use the same element `occurrenceConstraint` to store the mandatory constraint. The combination of a mandatory and uniqueness constraint is an occurrence constraint with comparator "equal" 1. Other numbers besides 1 can also be chosen, which corresponds to the frequency constraint in ORM. An occurrence constraint may be applied twice on the same path to specify ranges of occurrences.

Just as the external uniqueness constraint, mandatory constraints can also be placed on multiple paths. There are two constraints (with similar notation to external uniqueness). The `inclusive or constraint` states that the head term of the paths needs to play a role with at least one of the tail terms. The `exclusive or constraint` states that it needs to play a role with only one of the tail terms.

Set Comparison Constraints. In essence every path returns a set of instances. The *set comparison constraints* are used to specify comparison between sets of instances of two concepts in the ontology. The `setComparisonConstraint` element has a target and source set expression and an attribute `comparator` indicating how the two sets need to compare to each other. There are three types of set constraints in Ω-RIDL: (i) *subset constraint*: this constrains that the source set needs to be a subset of the target set; (ii) *set equality constraint*: source and target sets need to be equal; (iii) *set exclusion constraint*: source and target sets may not share any elements.

For example, the set of instances of Common Competence has to be equal to the union of the set of instances of Linguistic Competence and the Computer Competence.

```
"Common Competence" is equal to "Linguistic Competence" union "Computer Competence"

<setComparisonConstraint comparator="equality">      <headTerm>Linguistic Competence</headTerm>
  <sourceSetExpression>                                </lexonBasePath
    <lexonBasePath>                                  </sourceSetExpression>
      <context>HRM rules</context>                  <targetSetExpression>
      <headTerm>Common Competence</headTerm>           <lexonBasePath>
    </lexonBasePath>                                    <context>HRM rules</context>
  </sourceSetExpression>                                <headTerm>Computer Competence</headTerm>
  <targetSetExpression>                              </lexonBasePath>
    <setOperator operation="union">                  </targetSetExpression>
      <sourceSetExpression>                        </setOperator>
        <lexonBasePath>                            </targetSetExpression>
          <context>HRM rules</context>           </setComparisonConstraint>
```

The source and target expressions do not necessarily need to be paths. They can be a combination of sets by using `setOperators` as in the example. A `valueSet` can also be used in the set comparison constraint. In this set the values are listed explicitly.

Value Constraints. The *value constraint* constrains the possible values of the instances of a concept. They may only be placed on terms mapped to application

symbols (cfr. Sect. 3.6). For example, the Competence Level ranges from 0 to 5.
This is done by specifying an upper bound and a lower bound on a set expression:

```
"Competence Level" is (>= 0 and <= 5)
<valueConstraint comparator="greaterThanOrEqual" value="0">
  <setExpression>
    <lexonBasePath>
      <context>HRM rules</context>
      <headTerm>Competence</headTerm>
      <role>with</role> <coRole>of</coRole>
      <tailTerm>Competence Level</tailTerm>
    </lexonBasePath>
  </setExpression>
</valueConstraint>
<valueConstraint comparator="lessThanOrEqual" value="5">
  ...
</valueConstraint>
```

When we want to constrain the set of possible values the instances of a concept
may have, we use a subset constraint with respect to a value set.

3.4 Articulation

Articulation of terms in the lexon base assigns an explicit meaning to these terms.
It is a mapping of context-term pairs to a concept in the Concept Definition
Server. Every concept in the CDS contains glosses that explain the meaning of
the concept in natural language. This allows us to avoid the problem of synonyms
and homonyms of natural language in the ontology engineering process. The
CDS is currently based on Wordnet [8]. The articulation process is not part
of the commitment process in ontology engineering, but we need the concept
definitions in ontological commitments to assess which path constructions are
valid. The context term pairs listed with the concept definitions are treated as
equal concepts and thus paths may be constructed connecting them.

In the PoCeHRMOM project we use domain dictionaries such as O*NET
(linked to WordNet) to concept glosses. For example the context term pair ⟨HRM
rules, Speech Clarity⟩ is mapped to a concept with the gloss "The ability to speak
clearly so others can understand you.".

3.5 Interpretation

There are two kinds of interpretations in Ω-RIDL: relation interpretations and
lexical interpretations.

Relation Interpretations. In DOGMA, lexons at the level of the lexon base
are uninterpreted – with the obvious exception of articulation of terms as described above. "Real" interpretation is otherwise entirely delegated to the commitment layer. Also there are no rules on what the names of terms and roles
need to be. Thus we must indicate in Ω-RIDL how role co-role pairs need to be
interpreted. This is done by assigning special relations to a role co-role pair in a
specific context. In order to do that, we would need a domain canonical "relation library". Constructing such relation libraries is a topic of ongoing research,
outside of this paper's scope.

An ubiquitous relationship in ontologies is the subsumption relation. It is used for defining taxonomies and when it is interpreted by Ω-RIDL it is used among other things to decide which valid paths may be constructed. The T-Lex tool in addition exploits the buit-in nature of subsumption interpretation () to give such relationships a different visualization.

For example, the subsumption relationship is used to interpret the role pair of the lexon ⟨HRM rules, Common Competency, is-a, subsumes, Competency⟩. In Ω-RIDL, one writes the commitment as:

```
relation interpretation: in "HRM rules" is-a / subsumes is Subsumption
<specialRelation name="Subsumption"
  reference="http://www.wordnet-online.com/subsumption.shtml">
  <contextRoleCoRoleTriple>
    <context>HRM rules</context>
    <role>is-a</role> <coRole>subsumes</coRole>
  </contextRoleCoRoleTriple>
</specialRelation>
```

After this declaration every occurrence of "is-a" / "subsumes" roles in the context "HRM rules" will be interpreted as the subsumption relationship.

3.6 Lexical Interpretations

An application ontologically commits by mapping its application symbols to concepts in the ontology. In its current form Ω-RIDL supports committing relational databases and xml schemas to paths in the lexon base. For an in depth explanation of how relational databases are committed in Ω-RIDL we refer to the work of Verheyden et al [30].

The terms that are mapped to application symbols are treated as lexical object types (LOTs or value types in ORM).

3.7 Conceptual Queries

Conceptual queries query the committing database in terms of the ontology. When a database structure has been correctly committed to the ontology in the lexical interpretation part of Ω-RIDL, these conceptual queries can be translated to correct SQL statements. Ω-RIDL uses the same intuitive approach to query as path traversal as did the original RIDL query language [17]. These queries can easily be represented in the Ω-RIDL ML file as they are just set (path) expressions. For this reason it turns out that "sharing" conceptual queries is a convenient technique that contributes to a methodological process to develop the ontology itself [4]. For example the following query returns all the names of employees that have competences with Competence Level equal to 0.

```
list Name of Employee with Competence with "Competence Level" 0
<conceptualQuery name="low competence employees">
  <setExpression comparator="equal" value="0">
    <lexonBasePath> ... </lexonBasePath>
  </setExpression>
</conceptualQuery>
```

It is required that the first term, in the path used in the query, has been mapped to an application symbol (cfr. Sect. 3.6).

4   Related Work and Discussion

In Ω-RIDL we support all ORM constraints, except for the ring constraints. However, Ω-RIDL is much more flexible as it can place constraints on paths longer than one fact. In DOGMA Studio, T-Lex uses the ORM constraints as a graphical representation. When constraints are necessary on longer paths, Ω-RIDL constraints are entered textually. A lot of expressivity is allowed on constraints as they are placed on set expressions, which can be a combination of paths with set operators and even occurrences (instances). In Ω-RIDL constraints are applied by enforcing sets to return a specified set of values or enforcing a number of values that may be returned. Consider for example the following constraint:

`each Person is with at most one Address with Country with Name 'Belgium'`

Here we placed a constraint on a path longer than one directed lexon and even constrained it to be only applied at the instance level. It forbids persons to have more than one address in Belgium, but allows multiple addresses for every other country. This is only an example to demonstrate how semantically rich constraints can be specified with Ω-RIDL in a language that still appears very natural and can be read and understood by non-experts.

Our technique only allows binary relations, which is not a problem as N-ary relations can always be transformed (lossless) into binary relations [5]. Lisa-D [26] is a language for the PSM (Predicator Set Model) conceptual modeling technique, also based on NIAM. It does have support for N-ary relations and has a strong formalization, however at the cost of readability.

The ConQuer language [2] is the conceptual querying language for ORM. It is somewhat similar to RIDL as it is also based on the principle of traversing paths in the conceptual schema. It is not a language for specifying constraints, in that respect the authors deem ORM as sufficient.

5   Conclusions and Future Work

We introduced Ω-RIDL and its Ω-RIDL ML counterpart for specifying ontological commitments and conceptual queries in the DOGMA ontology engineering approach. It reuses most syntactical constructs with similar semantics of RIDL, a conceptual language on top of databases based on the NIAM modeling methodology. We introduced most semantic constraints by an example case study from the PoCeHRMOM project and briefly touched the subject of conceptual queries.

In the future we will formalize the language further and investigate how xml files can be conceptually queried by committing their schema. Also more constraints will be investigated from other languages such as OWL. Currently a prototype for an Ω-RIDL interpreter is being developed. Once this has been build more insights can be gained in how expressive Ω-RIDL is in conceptual querying. A compiler that translates Ω-RIDL into Ω-RIDL ML exists, but needs to be updated to support the new constructs. Finally we need to further integrate Ω-RIDL into our DOGMA Studio suite, as most of the statements can be generated by a CASE tool.

References

1. Berners-Lee, T., Hendler, J., Lassila, O.: The semantic web. Scientific American 284(5), 34–43 (2001)
2. Bloesch, A.C., Halpin, T.A.: Conceptual queries using ConQuer–II. In: ER 1997. Proc. of the 16th Int. Conf. on Conceptual Modeling, pp. 113–126. Springer, Heidelberg (1997)
3. De Leenheer, P., de Moor, A., Meersman, R.: Context dependency management in ontology engineering: a formal approach. In: Spaccapietra, S., Atzeni, P., Fages, F., Hacid, M.-S., Kifer, M., Mylopoulos, J., Pernici, B., Shvaiko, P., Trujillo, J., Zaihrayeu, I. (eds.) Journal on Data Semantics VIII. LNCS, vol. 4380, pp. 26–56. Springer, Heidelberg (2007)
4. de Moor, A., Leenheer, P.D., Meersman, R.: DOGMA-MESS. In: Schärfe, H., Hitzler, P., Øhrstrøm, P. (eds.) ICCS 2006. LNCS (LNAI), vol. 4068, pp. 189–203. Springer, Heidelberg (2006)
5. De Troyer, O.: On Data Schema Transformations. PhD thesis, Tilburg University, Netherlands (1993)
6. De Troyer, O., Meersman, R., Verlinden, P.: RIDL* on the CRIS Case: a Workbench for NIAM. In: Olle, T., Verrijn-Stuart, A., Bhabuta, L. (eds.) Information Systems Design Methodologies: Computerized Assistance during the Information Systems Life Cycle, pp. 375–459. Elsevier Science Publishers, Amsterdam (1988)
7. Falkenberg, E.D.: Concepts for modelling information. In: IFIP Working Conf. on Modelling in Data Base Management Systems, pp. 95–109 (1976)
8. Fellbaum, C.: WordNet: An Electronic Lexical Database. MIT Press, Cambridge (1998)
9. Gruber, T.R.: Towards principles for the design of ontologies used for knowledge sharing. In: Guarino, N., Poli, R. (eds.) Formal Ontology in Conceptual Analysis and Knowledge Representation, Kluwer Academic Publishers, Dordrecht (1993)
10. Guarino, N.: Formal ontology, conceptual analysis and knowledge representation. Int. J. Hum.-Comput. Stud. 43(5-6), 625–640 (1995)
11. Guarino, N.: Understanding, building and using ontologies. Int. J. Hum.-Comput. Stud. 46(2-3), 293–310 (1997)
12. Guarino, N.: Formal ontology and information systems. In: FOIS 1998. Int. Conf. On Formal Ontology In Information Systems, Trento, Italy, pp. 3–15. IOS Press, Amsterdam (1998)
13. Guarino, N., Welty, C.: Evaluating ontological decisions with ontoclean. Int. J. Hum.-Comput. Stud. 43, 625–640 (1995)
14. Halpin, T.A.: Information Modeling and Relational Databases: From Conceptual Analysis to Logical Design. Morgan Kaufmann, San Francisco (2001)
15. Jarrar, M., Meersman, R.: Formal ontology engineering in the DOGMA approach. In: Meersman, R., Tari, Z., et al. (eds.) CoopIS 2002, DOA 2002, and ODBASE 2002. LNCS, vol. 2519, pp. 1238–1254. Springer, Heidelberg (2002)
16. Jarrar, M., Demey, J., Meersman, R.: On using conceptual data modeling for ontology engineering. In: J. on Data Semantics. LNCS, vol. 2800, pp. 185–207. Springer, Heidelberg (2003)
17. Meersman, R.: Towards the very high level end user. In: Infotech State of the Art Report, Pergamon Press (1980)
18. Meersman, R.: The RIDL conceptual language. Research report, Int. Centre for Information Analysis Services, Control Data, Brussels (1982)

19. Meersman, R.: Semantic ontology tools in is design. In: Raś, Z.W., Skowron, A. (eds.) ISMIS 1999. LNCS, vol. 1609, pp. 30–45. Springer, Heidelberg (1999)
20. Meersman, R.: Ontologies and databases: More than a fleeting resemblance. In: d'Atri, A., Missikoff, M. (eds.) OES/SEO 2001 Rome Workshop, Luiss Publications, Missikoff (2001)
21. Meersman, R.: Web and ontologies: Playtime or business at the last frontier in computing? In: Proc. of the NSF-EU Workshop on Database and Information Systems Research for Semantic Web and Enterprises, pp. 61–67 (2002)
22. Mena, E., Kashyap, V., Sheth, A.P., Illarramendi, A.: OBSERVER: An approach for query processing in global information systems based on interoperation across pre-existing ontologies. In: Conf. on Cooperative Information Systems, pp. 14–25 (1996)
23. Spyns, P.: Object role modelling for ontology engineering in the DOGMA framework. In: Meersman, R., Tari, Z., Herrero, P. (eds.) OTM 2005 Workshops. LNCS, vol. 3762, pp. 710–719. Springer, Heidelberg (2005)
24. Spyns, P., Meersman, R., Jarrar, M.: Data modelling versus ontology engineering. Database Management and Information Systems 31(4), 12–17 (2002)
25. Tang, Y., Meersman, R.: Towards building semantic decision table with domain ontologies. In: Int. Conf. on Inf. Tech. and Management, pp. 14–22 (2007)
26. ter Hofstede, A.H.M., Proper, H.A., van der Weide, T.: A Conceptual Language for the Description and Manipulation of Complex Information Models. In: Gupta, G. (ed.) Seventeenth Annual Computer Science Conference: Gupta, G, vol. 16, pp. 157–167. University of Canterbury (1994)
27. Trog, D., Vereecken, J., Christiaens, S., De Leenheer, P., Meersman, R.: T-lex: A role-based ontology engineering tool. In: Meersman, R., Tari, Z., Herrero, P. (eds.) OTM 2006 Workshops. LNCS, vol. 4278, pp. 1191–1200. Springer, Heidelberg (2006)
28. van Griethuysen, J.J.: Concepts and terminology for the conceptual schema and the information base. Technical Report ISO/TC97/TR9007, International Standards Organization (1987)
29. Verheijen, G., van Bekkum, P.: NIAM, aN Information Analysis Method. In: Olle, T., Sol, H., Verrijn-Stuart, A. (eds.) IFIP Conf. on Comparative Review of Information Systems Methodologies, pp. 537–590. North-Holland, Amsterdam (1982)
30. Verheyden, P., De Bo, J., Meersman, R.: Semantically unlocking database content through ontology-based mediation. In: SWDB, pp. 109–126 (2004)
31. Zhao, G., Gao, Y., Meersman, R.: An ontology-based approach to business modelling. In: ICKEDS 2004. Proc. of the Int. Conf. of Knowledge Engineering and Decision Support, pp. 213–221 (2004)
32. Zhao, G., Meersman, R.: Architecting ontology for scalability and versatility. In: Meersman, R., Tari, Z. (eds.) On the Move to Meaningful Internet Systems 2005: CoopIS, DOA, and ODBASE. LNCS, vol. 3761, pp. 1164–1605. Springer, Heidelberg (2005)

Recovering Business Rules from Legacy Source Code for System Modernization

Erik Putrycz and Anatol W. Kark

Software Engineering Group, National Research Council Canada
{Erik.Putrycz,Anatol.Kark}@nrc-cnrc.gc.ca

Abstract. By using several reverse engineering tools and techniques, it is possible to extract business rules from the legacy source code that are easy to understand by the non-IT experts. These business rules can be used at different stages of system modernization. System maintainers can use the rules to locate in the code parts affected by a change in a rule. Business analysts can use those rules as means to aide understanding of the system at a business level. The extracted rules can serve as source of documentation and possible input for configuring a new system. This paper presents a novel approach for extracting business rules from legacy source code and application of the results at different stages of system modernization.

Keywords: Reverse engineering, Business Rules, Information Retrieval, System modernization.

1 Introduction

Governments and large corporations have a huge amount of the legacy software as part of their IT infrastructure. In 2006, 70% of all transaction systems were written in COBOL [1]. Understanding and discovery of business rules play a major role in the modernization of legacy software systems. According to a recent survey [2], about 51% of companies who have difficulties modernizing said that a major issue was the fact that "hard-coded and closed business rules" make it difficult to adapt their systems to new requirements and migrate to more modern environments.

This paper presents a process and the supporting tools for extracting business rules from legacy source code. Section 2 describes the context of legacy software system and its renovation, Section 3 details a process we used in analyzing and extracting business rules from COBOL code, while Section 4 presents and tool we developed and its possible use and early results of the analysis. Finally, Section 0 presents the conclusions and outlines direction of future research.

2 Legacy Software and System Renovation

According to [3], "Legacy Information Systems" (LIS) have the following characteristics:

- Usually run on obsolete hardware that is slow and expensive to maintain.

A. Paschke and Y. Biletskiy (Eds.): RuleML 2007, LNCS 4824, pp. 107–118, 2007.

- Software maintenance can also be expensive, because documentation and understanding of system details is often lacking and tracing faults is costly and time consuming.
- Lack of clean interfaces makes integrating LISs with other systems difficult.
- Are difficult, if not impossible, to extend.

We consider for this research large legacy "Enterprise applications" which were developed in-house and are critical to enterprise function. Those include pay, insurance, warehouse logistics and many similar systems.

The typical characteristics of these types of applications are:

- They handle and process a large amount of data.
- They have a large user base.
- They involve many business rules linked to legislation, processes and other legal matters.
- Many business rules are not accurate anymore and require external error and exception processing because of the difficulty to improve and update such system.
- The maintenance of the system often involves changes related to business rules.
- The source code is available.

By modernization, we mean here any process for evolving a system. This can be achieved by many different ways [4]. For example the legacy system can be replaced by a new one, or it can be interfaced with a new system. In any case, business rules in the legacy system play a major role. If the legacy system is being replaced, then the business rules are essential source of information for building the new system. If the legacy system is not phased out, then locating and analyzing the existing rules will help evolving and maintaining the legacy code.

In this context, we consider two main types of "business rules" [5]:

- From the business perspective, legislative rules, business practices and operational rules that are usually linked to natural language documents;
- From the Information System perspective, high level rules that are interpreted from the operational and legislative rules into executable rules that can be used in a system, possibly a business rules engine.

Two main classes of the stakeholders are involved with business rules:

- *Legacy system maintainers* fix bugs and implement new rules in the legacy system. They need to understand, not just the required changes, but the business rules they are affecting and the execution paths to a specific business rule.
- *Business analysts* must understand a context in which new requirements of a system exist and the effect those requirements will have on the existing rules.

3 Extracting Business Rules from Legacy Code

In the context of a system modernization, our objective is to provide tools that extract business rules from the legacy code, to support both a legacy system and a new system construction. Consequently, the requirements for extraction tools are:

- All extracted artifacts have to be traceable to the source code;
- Business rules must be at high level and understandable by all stakeholders involved, including business analysts.

In this section, we will focus on analyzing and extracting business rules from the COBOL programming language in a way that satisfies our stated objectives. The process and tools presented in this paper have been implemented for COBOL and were developed as part of two large system modernization projects in which we are currently involved. While COBOL is our source language we believe that the same method can be applied to other programming languages.

3.1 Business Rules Definition

According to the Business Rules Group, *business rules* are, from the business perspective, an obligation concerning conduct, action, practice, or procedure within a particular activity or sphere. From the information system perspective, a business rule is a statement that defines or constrains some aspect of the business [5]. The Semantics of Business Vocabulary and Business Rules (SBVR) standard [6] defines a standard vocabulary and a process for writing business rules. We will use here a small subset of SBVR called "production business rules" [7]. They have the following syntax [8]:

<conditions> *<actions>*
where:
<conditions>
1. consists of one or more Boolean expressions joined by 'logical' operators ("and", "or", "not")
2. must evaluate to a "true" state for the rule's actions to be considered for execution;
and

<actions>
1. consists of one or more Action expressions
2. requires the rule conditions to be satisfied (evaluate to true) before executing.

3.2 Business Rules Extraction and System Renovation

There has been lot of research done on system renovation techniques. Some of these techniques include [9]:

- reverse engineering,
- forward engineering,
- redocumentation,

- design recovery,
- restructuring,
- reengineering (or renovation).

According to [10], *reverse engineering* is concerned only with investigation of a system. *Redocumentation* is focused on making a semantically equivalent description at the same level of abstraction. Existing work on COBOL reverse engineering has focused on extracting diagrams [11], or translating it to other languages [12, 13]. A product called HotRod from Netron [13] aims to extract business rules from COBOL code. The granularity and extraction process are proprietary, but a user can define how many statements constitute a business rule. Additionally, to get more accurate results for each system, the algorithm used by HotRod should be tuned with the assistance of a COBOL expert. Once the tool has extracted business rules, a user needs to review the results and decide what do to; either use them for documentation purposes or use the results to restructure the code. If the results are to be used for documentation purpose, the results can be exported into a different documentation tool.

A product from Evolveware called S2T helps to automate the redocumentation process [14, 15] of legacy software. The tools analyses the source code using a knowledge engine that is "trained" for data discovery and knowledge extraction. The extracted information can also be an input for a new system. The business rules are focused on extracting the execution flow and can analyze end user input/output flows.

In the field of software maintenance, related work on redocumentating legacy systems has been done. In [16], the authors present a formal ontological representation that integrates Java source code and software documentation. Their tool parses Java code and using the AST they populate ontologies into a database. Using a custom natural language processor, they construct a second set of ontologies and then link it to the source code set. This process reduces the work required for redocumenting a system.

3.3 From COBOL to Business Rules

The structures of legacy COBOL and today's object oriented programs are radically different. COBOL has a limited and simple structure - each program contains a sequence of statements grouped into *paragraphs* which are executed sequentially. COBOL has only two forms of *branching*; one to execute external programs (CALL) and another to transfer the control flow to a paragraph (PERFORM). PERFORM is also used for iterations. Each program is associated with a single source file.

To extract useful information from the source code, it is necessary to find heuristics that separate setups, data transfers (usually from flat files) from business processing. To separate those two aspects, we focus on single statements that carry a business meaning such as calculations or branching since they most often represent high level processing.

In the case of branching (as defined above), Paragraph names and external programs can either be traced to documentation or the names themselves could possibly be self-explanatory. In code we have investigated, we found large number of paragraph names such as "CALCULATE-BASIC-PAY" which can be understood by a business analyst.

To make calculations easier to understand, it is necessary to translate the variable names, which more often are abbreviations or codes, into non-technical (their semantical) terms. We have encountered a large number of variable names such as CC35-CUR-APR or WS-PREV3-SUPN-LOW-MAX that require translation to the business terms.

Once calculations and branching are located, it is necessary to construct their *context*. Here, by context we mean all conditions in which a calculation or branching operation happens. This includes all IF statements under which an operation might be executed and also iteration information. This context also needs to be translated into non-technical terms. The conditions and statements are then converted into production business rules.

3.4 Knowledge Extraction Process

This business rules construction process is based on abstract syntax tree (AST) analysis. An abstract syntax tree is a tree, where the internal nodes are labeled by operators, and the leaf nodes represent the operands of the operators. This tree is obtained by parsing the source code. The parser required to construct the AST for this task had to be specifically designed to analyze and include all elements of the source code including comments. Many existing parsers are designed for other purposes such as syntax verification or transformation and they did not suit our objectives.

After creation of the AST, the business rules extraction is divided into two steps (Figure 1). First, we analyze the AST and construct a knowledge database. Once that database is constructed, we simplify the data collected (detailed in Section 0) and link the artifacts, wherever possible, to the existing documentation (Section 0). This is done by adding various semantic information, usually links connecting variable references, to the corresponding declarations.

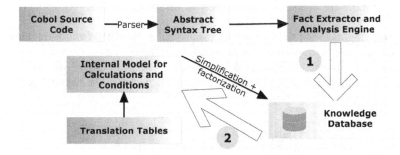

Fig. 1. Source code analysis process

The knowledge elements extracted from the abstract syntax tree are represented in Figure 2. The extracted elements are:

- production business rules: if *condition*, do *action*;
- conditions: one or more Boolean expressions with variables and constants joined by 'logical' operators;

- business rules: an action, possibly a condition and a comment from the source code;
- actions can be either branching statements or calculations with identifiers and constants;
- identifiers: variables used in conditions and calculations;
- Code blocks: they represent one or more lines of code in the original program;
- business rules dependencies: some calculations in other business rules executed before may affect the current calculation thus a business rules can be linked to one or other ones;
- loop and branching information: the paragraph in which the business rule is located may be called in a loop from another location in the program;

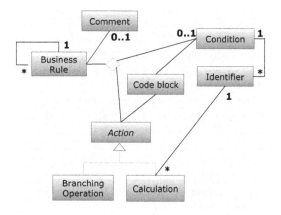

Fig. 2. UML Class diagram of the knowledge database

The following example illustrates the business rules extracted from a fragment of the COBOL code. It is important to notice that the *IF* conditions get unfolded and the two COMPUTE statements are treated as separate (not part of one IF statement).The extracted rules in this example have not yet been subjected to the semantical analysis.

COBOL code:

```
019872      IF M39PEN-CD = 18
019873          IF FUND2-PREV-YR-202 > 0
019874              COMPUTE FUND2-5B7-202 =
019875                      FUND2-5B7-202
019876                  + (FUND2-GROSS-202
019877                      * CORRCTNL-FACTOR)
019878          END-IF
019879          IF FUND3-2PYR-GROSS-202 > 0
019880              COMPUTE FUND3-5B7-202 =
019881                      FUND3-5B7-202
019882                  + (FUND3-GROSS-202
```

```
019883                          * CORRCTNL-FACTOR)
019884            END-IF
019885      END-IF
```

And extracted rules:

- BR-1: *if* M39PEN-CD = 18 and FUND2-PREV-YR-202 > 0, *calculate* FUND2-5B7-202 = FUND2-5B7-202 + (FUND2-GROSS-202 * CORRCTNL-FACTOR)

- BR-2: *if* M39PEN-CD = 18 and FUND3-2PYR-GROSS-202 > 0, *calculate* FUND3-5B7-202 = FUND3-5B7-202 + (FUND3-GROSS-202 * CORRCTNL-FACTOR)

4 Structuring and Navigating Extracted Business Rules

In order to enable a business analyst to understand the rules collected, all technical terms have to be translated into business terms. We are using several mechanisms for achieving this goal:

- Dedicated user interface for visualizing and navigating the collected data
- Integrating documentation
- Simplification of conditions in business rules

4.1 Business Rules Visualization

As a part of our work, we have built a rules visualization and browsing tool. Additionally, the tool can generate dynamically an execution graph for each of the rules. The tool takes as input the results of the source code analysis – the knowledge database.

The current implementation of the business rules extraction tools has been tested on 18 COBOL programs of a large legacy system representing about 630,000 lines of code. Our extraction tool found 9199 business rules, 4825 of them are calculation and 4374 branching operations. The complete generation of the knowledge database requires about one hour for analyzing the source code and generating business rules. However, the generation time has not been an issue since the database doesn't need to be updated instantly.

Figure 3 shows the main elements of the graphical user interface of the tool. The left panel (item 1) is used to navigate the business rules in different ways. The right panel (items 2 and 3) displays the information about a selected element in the left panel. In this case, Figure 3 displays the details of a business rule. Those details include the context of the rule (current paragraph and current program in item 2), conditions and the action itself (item 3). All variable names are translated using the strategy detailed later. Using the type information, we color each element type (identifiers, constants, etc.) differently. When the action is a calculation, we convert it to an equation and render it in a form that business analysts are more used to.

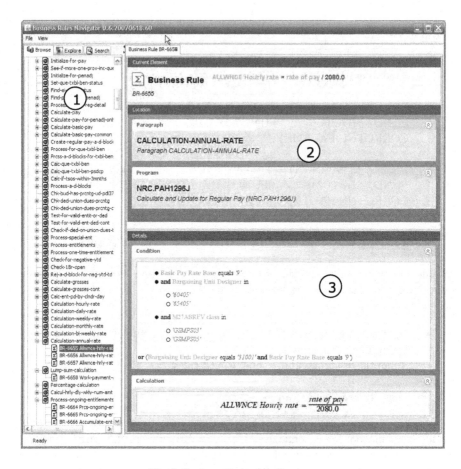

Fig. 3. Business Rule visualization

In the knowledge database, the conditions and actions of business rules are associated with *code blocks* (Figure 2). A code block can be any (disjoint) section of the source code. Figure 4 shows highlighting used to trace the conditions and actions of a business rule to the source code. When the user selects a business rule and opens the source code view tab of the tool, the code sections corresponding to the conditions and the action are highlighted. This feature is targeted at system maintainers who need to update and modify the business rules.

4.2 Integrating Documentation

Although legacy systems are often not well documented, there usually is some documentation of the data processed, including at least the names of the fields and their descriptions. Our objective is to link that existing documentation with the business rules.

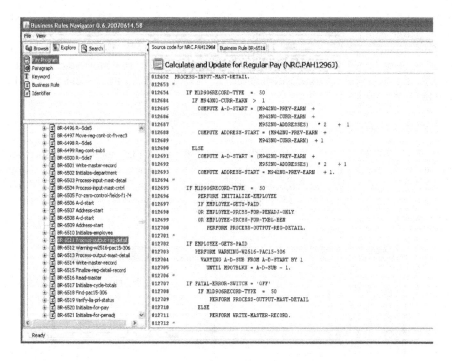

Fig. 4. Business rules and source code linking

One way to achieve this is through variables names. For each variable, we try different strategies to translate their technical names. At first, if the variable is directly linked to a data field (usually a field of the so called "master file") then we use that field description. Otherwise, we look in the AST and search for variable assignments and copies that can link one variable to a documented one. This scenario happens often in COBOL because the developers use temporary variables to do all processing after reading and before writing the data. Finally, if we cannot trace the variable to a documented field, we use a translation table with abbreviations and their common meaning. This process ensures that the majority of identifiers are presented with their business semantics.

4.3   Simplification of Conditions in Business Rules

The condition of a business rule is based on all of the *IF* statements which have to be true for an action to be executed. The sum of all these conditions can easily result in a long and complex expression difficult to understand. To greatly reduce the size and complexity of the expressions, we use - according to the user's preferences - either a conjunctive normal form (CNF) or a disjunctive normal form (DNF).

In addition to the normal form, we use type inference to match constants used in the expressions with data tables. A custom type can be assigned to variables in our knowledge database which then can be used by the expression builder to assign these types to constants. The user interface translates these constants using custom types and existing data files or database tables.

For instance the following condition "ELEMENT62-ENT-DED-CD = '304' OR '305'" can be translated to "Entitlement Deduction in (overpayment, rec o/p extra duty)". By inferring the type "deduction" to '304' and '305', their meaning can be found in an external data table.

4.4 Navigation Modes

Recent studies [17] show how that a flexible navigation is a key element for software maintainers. According to their findings, the development tools have to show dependencies, allow navigation among them and provide an easy way to judge of the relevance of information. It is also necessary to provide means to collect the relevant information.

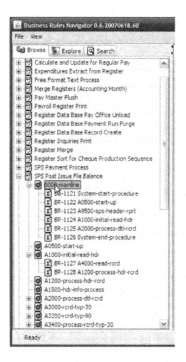 

Fig. 5. Basic navigation   **Fig. 6.** System wide navigation

For these purposes, three different types of navigation of the extracted information are available:

- Navigation in a tree with the following hierarchy: the different COBOL programs, the paragraphs and the business rules inside each paragraph (Figure 5);

- Navigation through the whole system using keywords attached to business rules, identifiers, paragraphs and comments; Figure 6 displays an example of all business rules with the keyword "TAX" attached to them.
- Custom navigation using tags: users can create and add tags to any element and then navigate by tag through these elements.

4.5 Execution Graph Visualization

The objective of the execution graph visualization is to help understanding in which context one business rule might be executed. By context, we mean here all the executions paths in a program that can lead to the execution of the rule. Figure 7 shows an example of the different execution paths for the business rule BR-6501. The rule is at the bottom of this example graph. The remaining nodes in the graph are paragraphs preceding the execution of BR-6501. The edges represent the ordering. A label on the edge represents a condition under which the path is taken.

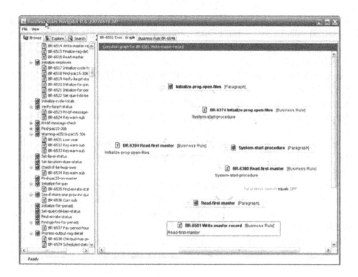

Fig. 7. Execution graph visualization

5 Conclusions

This paper presents means of helping a modernization process by extracting business rules from legacy source code. The novelty in this research is that the output is targeted at business analysts. This means that the business rules must be translated into non-technical terms. The extraction of rules from source code is divided into two main steps. First, the source code is parsed into an abstract syntax tree to locate the calculations and other elements of business rules. Second, a context is constructed for this information to transform it into a form which is easier to understand. This is done by analyzing all the conditions of a business rule and translating them to non-technical terms. Given the large amount of information collected, we implemented a dedicated interface to navigate the business rules. This process and tools are currently being used in two large projects for both modernization and legacy code maintenance.

Early feedback from business analysts has been extremely positive about the value and understandability of the information extracted. We plan to evaluate the effectiveness of the business rules extraction process by studying how business analysts use the results. In addition, we are looking at integrating natural language techniques such as co-location for improving the translation of identifiers.

References

[1] Krill, P.: The future's bright.. the future's COBOL (2006)
[2] Software, A.G.: Customer survey report: Legacy modernization. Technical report (2007)
[3] Bisbal, J., Lawless, D., Wu, B., Grimson, J.: Legacy information systems: issues and directions. Software 16(5), 103–111 (1999)
[4] Koskinen, J., Ahonen, J., Sivula, H., Tilus, T., Lintinen, H., Kankaanpaa, I.: Software modernization decision criteria: An empirical study. In: CSMR 2005. Software Maintenance and Reengineering, 2005. Ninth European Conference, March 21-23, pp. 324–331 (2005)
[5] Business Rules Group: What is a business rule? (2007), http://www. Usinessrulesroup. rg/defnbrg.shtml
[6] OMG: Semantics of Business Vocabulary and Business Rules (SBVR)
[7] Hendryx, S.: SBVR and MDA: Architecture. Business Rules Journal 6(11) (November 2005)
[8] OMG: Production Rule Representation Request For Proposal (2003)
[9] Chikofsky II, E.J., C, J.H.: Reverse engineering and design recovery: A taxonomy. IEEE Software, 13–17 (January 1990)
[10] Brand, M.G.J., van den, P.K., Verhoef, C.: Reverse engineering and system renovation: an annotated bibliography. ACM Software Engineering Notes 22(1), 42–57 (1997)
[11] Edwards, H.M., Munro, M.: RECAST: reverse engineering from COBOL to SSADM specification, 499–508 (April 1993)
[12] Gray, R., Bickmore, T., Williams, S.: Reengineering Cobol systems to Ada. Technical report, InVision Software Reengineering, Software Technology Center, Lockheed Palo Alto Laboratories (1995)
[13] Arranga, E.: COBOL tools: overview and taxonomy. Software 17(2), 59–69 (2000)
[14] EvolveWare Corporation: S2T legacy documentation tools (2007), http://www. volveware.com
[15] EvolveWare Corporation: Evolveware's S2T technology, a detailed overview (2007), http://www.evolveware.com
[16] Witte, R., Zhang, Y., Rilling, J.: Empowering software maintainers with semantic web technologies. In: Proceedings of the 4th European Semantic Web Conference (2007)
[17] Ko, A., Myers, B., Coblenz, M., Aung, H.: An exploratory study of how developers seek, relate, and collect relevant information during software maintenance tasks. Software Engineering, IEEE Transactions on 32(12), 971–987 (2006)

An Approach for Bridging the Gap Between Business Rules and the Semantic Web

Birgit Demuth and Hans-Bernhard Liebau

Technische Universität Dresden, Department of Computer Science
{birgit.demuth,hans-bernhard.liebau}@inf.tu-dresden.de

Abstract. Business rules should improve the human communication inside of an enterprise or between business partners and must be therefore independent of implementations in IT systems. As a long-term goal, business rules should be guaranteed by all IT applications of an enterprise. A first step to define and to standardize what business rules are is an OMG initiative to specify a metamodel for business rules and the vocabulary on which business rules are defined. The result of OMG's effort is the SBVR (Semantics of Business Vocabulary and Business Rules) specification that we took as starting point of our investigations to automate business rules. There are multiple ways for transforming business rules. In this paper we show how SBVR based vocabulary and rules can be translated by model transformation chains into Semantic Web Languages. In our approach we use OWL and R2ML (REWERSE Rule Markup Language). Both are languages with a high potential for a broad usage in future rule-based applications.

1 Introduction

Business vocabulary and business rules (BVBR) represent (a part of) the static view of business people on their business. BVBR should improve the human communication inside of an enterprise or between business partners and must be therefore independent of implementations in IT systems. In the context of OMG's Model Driven Architecture (MDA), the formulation of BVBR is considered as a Computer Independent Model (CIM). In the long run, business rules should be guaranteed by all IT applications of an enterprise. This needs a technology to transform the CIM of BVBR to IT applications. In the words of the MDA terminology, we have to transform in a first step the CIM into a PIM (Platform Independent Model). In this paper, we meet this challenge and call the transformation technology of rule models at different abstraction levels to the point of executable rules *Model-Driven Integrity Engineering*. As a starting point of our investigations to automate business rules, we use the OMG specification SBVR (Semantics of Business Vocabulary and Business Rules) [3]. SBVR is independent of a specific natural language. Instead it provides a metamodel for business rules and the vocabulary on which business rules are defined. Concrete syntax SBVR examples are in English, but languages with different sentence ordering have been taken into account. The vocabulary basically consists of terms,

A. Paschke and Y. Biletskiy (Eds.): RuleML 2007, LNCS 4824, pp. 119–133, 2007.

definitions for concepts, and relations between concepts. Rules are based on the terminology defined in the vocabulary. Their underlying semantics is defined by statements in predicate logic.

Until now, the SBVR specification says all but nothing about how to derive a PIM model and rule expressions from SBVR vocabulary and rule statements. However, multiple ways for transforming business rules are being discussed. One option is the mapping of SBVR vocabulary and rules to a UML class model including OCL (Object Constraint Language [1]) expressions (especially invariants and derived expressions) specifying some types of business rules [4]. OCL could also serve as condition language in a PRR rule model. PRR (Production Rule Representations, [2]) is an OMG specification that is proposed for the specification of production rules at the PIM level. Production rules are commonly used to represent business rules in IT systems and are supported by a number of rule engine respectively rule management system vendors. A further discussed option automating SBVR business rules is based on the similarity of SBVR and semantic web artifacts. Both specify *semantics* or meaning for example by *rules*, *terms* and *concepts*. An immediately viewable difference consists in the syntax. As SBVR proposes human readable descriptions to specify meaning, Semantic Web uses machine readable descriptions and is therefore adequate for a PIM model [5].

In this paper we investigate the second option and show how SBVR based rules can be translated by a model transformation chain into Semantic Web Languages. As transformation language we use the QVT (Query/View/Transformation) language from the OMG [6]. All transformations are metamodel based. SBVR has a Vocabulary-to-MOF/XMI Mapping Rule Set to produce a MOF [10] model and XML schema from the SBVR vocabulary. As target languages we use in our approach OWL and R2ML (REWERSE Rule Markup Language). OWL [9] is the well-known Web Ontology Language that can be used to describe classes and relations between them that are inherent in Web documents and applications. In order to describe OWL models within the MOF context, we use the Ontology Definition Metamodel (ODM) [11] from the OMG. R2ML [7] is already designed as a MOF-based general rule markup language serving as a pivotal metamodel for the interchange of rules between different rule engines. The language is one of the well-known W3C proposals for a future Rule Interchange Format (RIF) [8].

Our design considerations for the choice of the above mentioned languages including a small example are explained in Section 2. Section 3 introduces at first an intermediate metamodel to simplify and improve the efficiency of the SBVR-to-OWL/R2ML transformation. Then the necessary metamodel-based transformations are presented from a conceptual point of view. In Section 4 we report about first experimental results with prototyping of the defined transformations. In the closing section we summarize our results and submit a vision for future use and automation of SBVR-based business rule languages.

2   Design Consideration

2.1   SBVR as Source Language

The basic concepts of SBVR are noun concepts, roles and fact types (all of them describing vocabularies) as well as rules (in the sense of integrity rules) whereas all the information is given by facts instantiated on the basis of fact types again. Fact types contain one, two or more roles. For instance, the binary fact type "concept has designation" of the SBVR specification itself contains the two roles "concept" and "designation". Rules are special facts that have still another relation to fact types. They are propositions using fact types on the same modeling level to express constraints for instances of these fact types. There are two kinds of rule modalities; the alethic ones, e.g. a necessity, and the deontic ones, e.g. an obligation. Let us look at a small example. It is a fragment of an SBVR-based domain model following [12], presented in a similar schema as the SBVR-structured English [3] does.

Vocabulary:

```
employee            ...
manager
        Definition      employee that manages others
        General concept employee
        Concept Type    role
        Synonym         supervisor
employee(1) is under employee(2)
        Synonymous Form employee(2) over employee(1)
   ...
```

Rules:

```
MC No Selfmanagement
        Rule Statement  It is prohibited that an employee
                        is a manager over the employee.
   ...
```

The example describes an employee hierarchy in an enterprise as well as a deontic rule claiming that no employee shall manage her/himself. The vocabulary contains noun concepts and fact types. In our example, we have three items ("employee", "manager" and "employee is under employee"). The rule statement of the above example is based on two fact types, "thing is thing" being an elementary concept of SBVR and describing equality of things, as well as the former described domain fact type "employee is under employee" represented by its synonymous form "employee over employee". In a logic-like notation, the rule MC No Selfmanagement could be written as follows:

$$\Box_{deo} \neg \exists m \in \text{manager} \; \exists e \in \text{employee}(\text{thing is thing}(e,m) \mid \text{employee over employee}(m,e))$$

Let us look at the above mentioned domain fact type coded in an XMI serialized way [12]:

```
<xmi...>
  <esbvr:Thing
      xmi:id="employee-is-under-employee-fact-type"/>
  <esbvr:Thing
      xmi:id="employee-is-under-employee-sentential-form"/>
  <esbvr:Thing
      xmi:id="employee-over-employee-noun-form"/>
  ...
  <esbvr:Extension
      xmi:id="employee-is-under-employee-roles"
      element="employee-is-under-employee-role1..."
  ...
  <esbvr:Text xmi:id="t-employee-is-under-employee"
      value="employee is under employee"/>
  ...
  <sbvr:fact-type-has-role
      fact-type="employee-is-under-employee-fact-type"
      role="employee-is-under-employee-roles"/>
  ...
  <sbvr:fact-type-has-form-of-expression
      concept="employee-is-under-employee-fact-type"
      form-of-expression="employee-is-under-employee-sentential-form"/>

  <sbvr:fact-type-has-form-of-expression
      concept="employee-is-under-employee-fact-type"
      form-of-expression="employee-over-employee-noun-form"/>
  ...
</xmi:xmi>
```

The *Essential SBVR* is a special vocabulary and rule set of SBVR for its representation within the Essential MOF (EMOF) context [10]. It enables to create Things, Extensions, and auxiliary concepts for representing texts or integers (see the upper area in Fig. 1). All other information are instantiated by facts based on SBVR fact types inheriting from the Fact class. Some common SBVR fact types are depicted in the lower area in Fig. 1. Regarding their names, the figure is an EMF [13] representation of SBVR's MOF metamodel. For instance, the fact type "concept has designation" is represented as class named ConceptHasDesignation. In the EMOF context there are also so-called instantiation fact types being special unary fact types (containing only one role) that classify Thing-s if there are no sufficient information inside other fact types. All fact types directly inherit from the Fact or Instantiation class so that the hierarchy depth from there is just one. The class tree is very wide. All in all there are around 470 fact types represented as EMOF classes. In our practical considerations we just focused on a selected subset of fact types applied in a domain model example.

In SBVR the open world assumption is assumed by default. Moreover fact types can be explicitly declared as "closed". Then all the facts of such a type are assumed as explicitly known within the domain. The primary SBVR use of MOF is

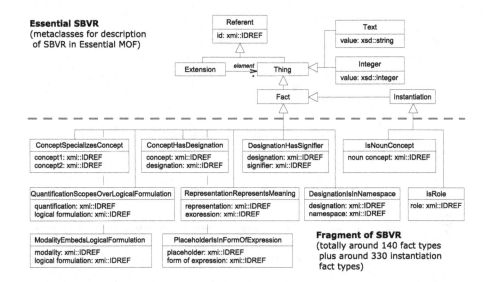

Fig. 1. Fragment of the SBVR MOF metamodel (subset of fact types and their relation to Essential SBVR; EMF representation)

interchange of vocabularies and rules between tools (see Annex K.2 in [3]). In contrast to the fact-oriented approach [14] which SBVR is based on, EMOF corresponds more closely to many facilities found in object oriented programming languages [10]. For modeling in SBVR, the semantic analysis facilities of MOF are not exploited completely. Among others, there is no MOF representation of the SBVR noun concept hierarchy, with the exception of the Essential SBVR concepts (see Fig. 1). For instance, it is not mapped to MOF that the SBVR concept "role" is a subconcept of the SBVR concept "noun concept" (see Fig. 8.1 in [3]).

2.2   OWL as Target Vocabulary Language

For mapping SBVR vocabularies into the Semantic Web we chose the Web Ontology Language (OWL) [9] as target language. SBVR as well as OWL enable a logic based description of terms, their classifications and their relations to each other. Both languages are based on the open world assumption [5]. OWL is the currently established standard that most expressively describes ontologies and is supported by multiple tools. This helps to preserve as much semantics as possible and allows prototyping to prove our aproach. We chose the OWL-Full dialect. So we are able to describe domain model elements both as classes and as individuals (see Section 3.3) which is not possible by other OWL dialects. By omitting the mapping of some information the model is easily reducible into an OWL-DL based model. This can be useful if only reasoning is required. In order to describe OWL models within the MOF context, we use the Ontology Definition Metamodel (ODM) [11], partially depicted in Fig. 2.

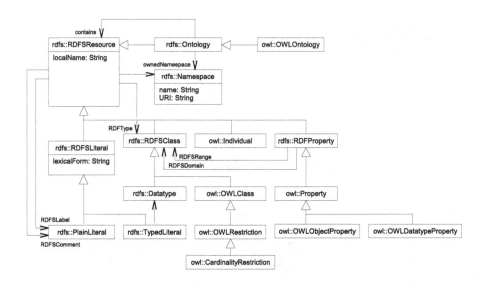

Fig. 2. Fragment of the Ontology Definition Metamodel (ODM) [11]

2.3   R2ML as Target Rule Language

SBVR rules have to be mapped into a rule language. Since SBVR is extremely expressive we have to choose a language with an expressive power as high as possible regarding the scope of SBVR. R2ML [7] is an adequate rule language in the Semantic Web provided by the REWERSE project [15]. It allows specification of derivation rules, production rules and reaction rules. Furthermore, it is possible to formulate deontic and alethic modalities for integrity rules. The vocabulary which the rules build on can be described by an external vocabulary language such as OWL. Fig. 3 depicts a fragment of the R2ML metamodel whereas only the integrity rules are represented here. In contrast to SBVR only one modality per rule is describable by R2ML. Furthermore, modality at first, quantifications at second and remaining formulations inside of these ones are to be set in this strict embedding order. The following code fragment of an integrity rule should make this fact clearer. The logicalFormula, e.g. for our above rule MC No Selfmanagement has still to be included. The equivalent code we consider in Section 3.4.

```
<r2ml:RuleBase...>
  <r2ml:ruleSet xsi:type="r2ml:IntegrityRuleSet">
    <r2ml:integrityRule xsi:type=...>
      <!-- e.g., type="r2ml:DeonticIntegrityRule"
           for an obligation claim -->
      <r2ml:constraint>
        <r2ml:quantifiedFormula xsi:type=...>
          <!-- e.g., type="r2ml:UniversallyQuantifiedFormula"
               for an universal quantification -->
          <r2ml:objectVariable classID=... name=.../>
```

```
<!-- the range of the variable is given by the classID
        attribute assigning, e.g., some OWL class id -->
<r2ml:logicalFormula xsi:type=...>
  ... <!-- atomic formulas, among others,
           see supplementary example in section 3.4 -->
</r2ml:logicalFormula/>
    </r2ml:quantifiedFormula>
  </r2ml:constraint>
</r2ml:integrityRule>
  </r2ml:ruleSet>
</r2ml:RuleBase>
```

Another restriction of R2ML is that only necessity and obligation modalities are describable, but no possibility or permission modalities. Therewith R2ML rule expression power is significantly less than the expression power of SBVR. In some cases it will be possible to transform a rule into a normalized one that is R2ML compliant. But this remains out of our consideration.

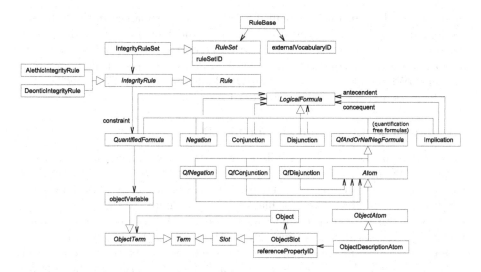

Fig. 3. Fragment of R2ML 0.4 metamodel (according to [7])

3   Model Transformation Chain

If we describe SBVR models as explained above in MOF/XMI [3] each fact type is encapsulated in a separate MOF class. Therewith the majority of transformation code becomes very complex. Furthermore the performance suffers during the processing of requests to fact-oriented models due to the search for scattered information in the SBVR model. Long chains of subrequests arise. Therefore the whole transformation is divided into multiple steps (trafo 1, trafo 2, and trafo 3 in Fig. 4). Trafo 1 describes the transformation from the source metamodel

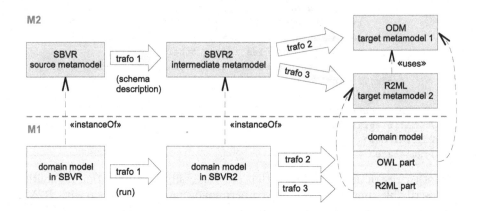

Fig. 4. Transformation chain

in an intermediate metamodel, called SBVR2 that is explained in the following section. It simplifies the transformation code and increases the performance. Therefore all further transformations are based on the intermediate metamodel. Trafo 2 describes the transformation of the vocabulary part of the domain model into an OWL ontology based on the ODM metamodel. Trafo 3 specifies the transformation of the rule part into R2ML.

3.1   Intermediate Metamodel for SBVR

Being aware of some loss in flexibility and expressive power, we designed a further MOF metamodel (SBVR2) exploiting more MOF features (Fig. 5). It corresponds to the original one (Fig. 1). SBVR2 contains a real MOF based representation of the SBVR concept hierarchy and is limited to the representation of unary and binary fact types on the metamodel level. But this is unproblematic because the specification contains only one fact type with an arity of n<2 that is, however, irrelevant. Of course, the advantages of the fact-oriented approach that SBVR originally uses (see Annex K.2 in [3]) are diminished by the intermediate metamodel. For instance, some SBVR structure features, like multidimensional class hierarchies, have to be represented in other ways (see [16]). But the presented metamodel is advantageous for the transformation into OWL/R2ML because it bundles all the facts for each original `esbvr:Thing` resource in an appropriately typed SBVR2 instance. The SBVR2 concepts, corresponding to SBVR noun concepts, allow easy and quick access to all relevant facts in the form of MOF class references, corresponding to the SBVR binary fact types.

3.2   Transformation into the SBVR2 Metamodel (Trafo 1)

The transformation can be done efficiently as explained above. Each binary SBVR fact type (see subclasses of `Fact` in Fig. 1) is transformed in two steps: At first, an SBVR2 instance is constructed for each role contained in the fact

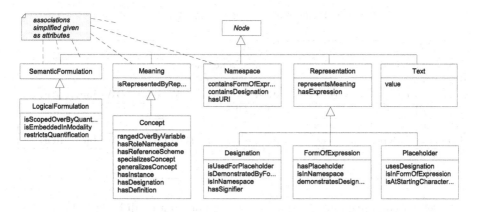

Fig. 5. Fragment of the intermediate metamodel (SBVR2 metamodel)

type if that instance has not been created yet (see classes in Fig. 5). Thereafter both instances are linked to each other using object references (see attributes in Fig. 5). For instance, if we have to map a fact, based on the SBVR fact type `FactTypeHasFormOfExpression`, we will map to the SBVR2 concepts `FactType` and `FormOfExpression`. The associated SBVR2/XMI-based code for the "employee is under employee" domain fact type could look like follows:

```
<sbvr2:DocumentRoot ...>
  <contains
      xsi:type="sbvr2:FactType"
      id="employee-is-under-employee-fact-type"
      hasFormOfExpression="employee-is-under-employee-sentential-form
          employee-over-employee-noun-form ..."
      hasRole="employee-is-under-employee-role1 ..."
      .../>
</sbvr2:DocumentRoot>
```

The right processing order of the fact types to be treated during the transformation is of big significance because each new SBVR2 instance has to be classified to its most specialized class. MOF classes can not easily change in more specialized ones during the transformation run-time. For instance, if we have to map a fact based on the SBVR fact type "concept has designation", we should already have mapped all facts based on the fact type "fact type has form of expression". The related SBVR noun concept "fact type" of the second fact type is a subconcept of the related SBVR noun concept "concept" of the first fact type. In SBVR2, objects of both noun concepts could be identical and are described as MOF classes of different specialization levels. That's why the second one has to be favored for the construction order.

3.3   Transformation into the ODM Metamodel (Trafo 2)

Table 1 shows an overview of the transformation from SBVR2 into the ODM metamodel from a conceptual point of view. Both `NounConcept` and `FactType`

Table 1. Basic mapping rules from SBVR2 to OWL

SBVR2	ODM
VocabularyNamespace	RDFSNamespace
RoleNamespace	RDFSNamespace
NounConcept / Designation	RDFSResource (OWLClass, OWLIndividual)
Role / Placeholder	OWLObjectProperty
IndividualConcept / Designation	OWLIndividual
FactType / Representation	RDFSResource (OWLClass, OWLIndividual)
Placeholder (additional information)	OWLIndividual

elements are mapped into `RDFSResource` elements first of all being `OWLClass` elements to the purpose of further instantiations toward the domain level. For each synonymous representation one target element is constructed. The synonymy is managed by the built-in `OWLEquivalentClass` property, setting equivalent OWL classes. In a similar way, roles of fact types are mapped into `OWLObjectProperty` elements whereas the domains of the OWL object properties are the owning fact types and the ranges correspond to generalizations of the roles inside the fact types (not given in the table). Placeholder elements declare some information about role representations inside any fact type representation like, e.g., the starting character position and the assigned role. If we wish to map such information, the `FactTypes` and `NounConcepts` have also to be addressed as `OWLIndividual` elements. To use OWL resources both as classes and as individuals we need the OWL-Full dialect. The detailed transformation is presented in [16]. To improve the human readability, especially of the RDF/XML output code of OWL, it seemed beneficially to introduce some stereotypes like "fact_type" and "noun_concept" both being OWL classes as well as subclasses of `OWLClass`. They are practically defined on the domain model level but describe a virtual meta level. An example of output code defining and using these stereotypes is given at the end of this subsection. From another point of view, stereotypes could be confusing in subsequent transformations.

In order to demonstrate the transformation into OWL, we look, e.g., into domain elements typed as noun concepts including their designations (see Fig. 6). The notation follows slightly the declarative graphical noation of OMG's QVT specification [6]. In alteration of that the arrow in the middle has only one head directed to the target model. Furthermore a `new` tag on top of a class box denotes classes to be created during the current mapping. Each `Designation` SBVR2 element has to be transformed into an `OWL` class from the stereotype `noun_concept`. The `id` attribute of the `Designation` is mapped to the `localName` attribute of the `OWLClass`. The `value` attribute of the owned signifier (`hasSignifier`) is allocated to the `RDFSLabel` attribute. The `Definition` is allocated to the `RDFSComment` attribute. The model transformation is completed by the assignment of the `Namespace`. A group of `Designations` representing the same noun

noun conceptToOWLClass

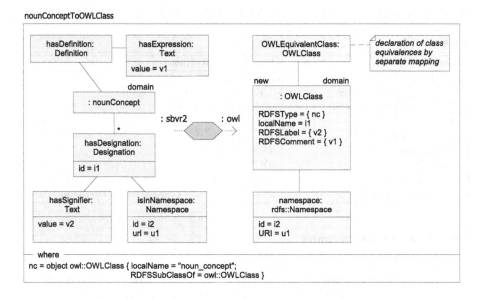

Fig. 6. Mapping SBVR noun concepts from SBVR2 to OWL

concept corresponds to a group of OWL classes that are linked by the `OWLEquivalentClass` attribute. The associated RDF/XML code is illustrated by the noun concept `employee` as follows:

```
<rdf:RDF ...>
  <owl:Ontology ...#domainModel"/>
  <!-- profile definition, could be externalized -->
  <owl:Class rdf:about="...sbvrProfile#noun_concept">
    <rdfs:subClassOf rdf:resource="...owl#Class"/>
  </owl:Class>
  ...
  <!-- real domain model elements -->
  <sbvrProfile:noun_concept rdf:about="...domainModel#employee-term">
    <rdfs:comment>person working for the company</rdfs:comment>
    <rdfs:label>employee</rdfs:label>
  </sbvrProfile:noun_concept>
  ...
</rdf:RDF>
```

3.4   Transformation into the R2ML Metamodel (Trafo 3)

Table 2 shows an overview of a fragment of the transformation from SBVR2 into the R2ML metamodel we treated in our work, from a conceptual point of view. Within the SBVR context only derivation and integrity rules are relevant. A recursive mapping algorithm is used to create the embedded structure of R2ML: First the modality is mapped, inside of this the quantifications are

Table 2. Basic mapping rules from SBVR2 to R2ML (fragment treated in our work)

SBVR2	R2ML
ObligationClaim	DeonticIntegrityRule
Quantification: - UniversalQuantification - ExistentialQuantification - etc.	QuantifiedFormula: - UniversallyQuantifiedFormula - ExistentiallyQuantifiedFormula - etc.
logical formulation: - conjunction - disjunction - logical negation - atomic formulation	logicalFormula: - Conjunction - Disjunction - StrongNegation - Atom (especially ObjectDescriptionAtom)

AtomicFormulationToAtomicFormula

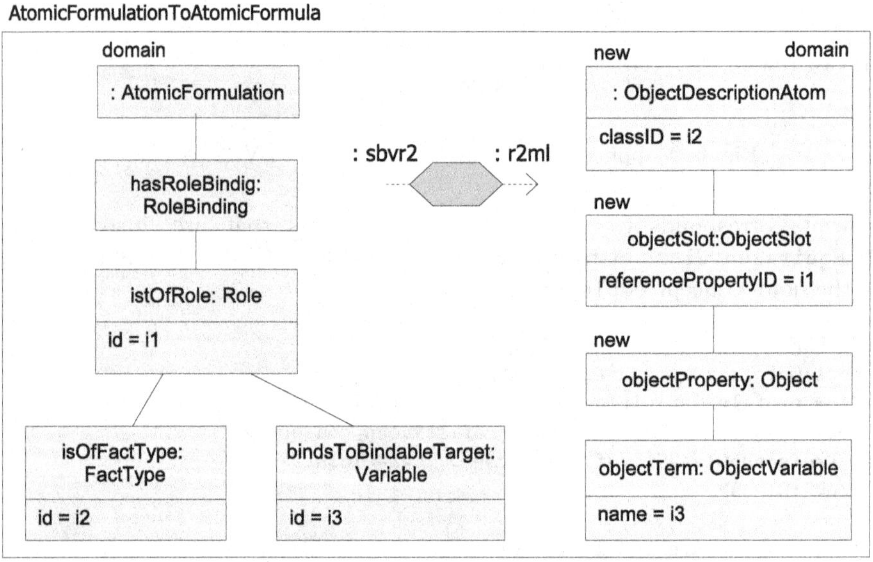

Fig. 7. Mapping SBVR atomic formulations from SBVR2 to R2ML

mapped, and inside of that the remaining logical formulations are mapped (see code fragment in Section 2.3). The inner expressions include the atomic formulations whose mapping is exemplarily presented in Fig. 7. In general an atomic formulation is based on a fact type that has one or several roles. It maps to an ObjectDescriptionAtom element whose classID attribute contains the name of the mapped fact type. The bounded roles of the atomic formulation are mapped to ObjectSlot classes contained in the ObjectDescriptionAtom class and containing the name of an introduced ObjectVariable. The variables are usually introduced by quantifications (see also code fragment in Section 2.3). Finally the variables are mapped, that is, inside the target model, the R2ML objectTerm properties, contained by the relevant `objectSlots`, are linked to the appropriate R2ML object variables that have been previously introduced by the outer quantified formulas (more detailed described in [16]). Elaborating the embedded `LogicalFormula` element of the `quantifiedFormula` tag, one of the atomic formulations, "employee over employee (m,e)", of the example rule described in Section 2.1 would be mapped as follows:

```
<!-- see code example in section 2.3 for embedding code -->
  <r2ml:LogicalFormula xsi:type="r2ml:ObjectDescriptionAtom"
     r2ml:classID="employee-is-under-employee-fact-type">
    <r2ml:ObjectSlot r2ml:referencePropertyID="employee-is-under-employee-role1">
      <r2ml:object>
        <r2ml:ObjectTerm xsi:type="r2ml:ObjectVariable" r2ml:name="e"/>
      </r2ml:object>
    </r2ml:ObjectSlot>
    <r2ml:ObjectSlot r2ml:referencePropertyID="employee-is-under-employee-role2">
      <r2ml:object>
        <r2ml:ObjectTerm xsi:type="r2ml:ObjectVariable" r2ml:name="m"/>
      </r2ml:object>
    </r2ml:ObjectSlot>
  </r2ml:LogicalFormula>
```

4 Prototyping

For a prototypical implementation we used the Borland Together Release 2 for Eclipse tool. We implemented the transformations in the Together version of the QVT operational language. The metamodel descriptions had to be specified by means of the Eclipse Modeling Framework (EMF) [13] on which Together for Eclipse is built on. The four EMF metamodels were created in different ways. The SBVR metamodel was imported from some `.xsd` files provided by the OMG. The SBVR2 metamodel was manually created by means of the Ecore editor of the EMF. An implementation of the OMG's Ontology Definition Metamodel (ODM) in EMF is provided by IBM as an Eclipse plugin open-source project (EODM) [17]. And the R2ML metamodel was generated again from some `.xsd` and related module files provided by the REWERSE project. To implement the above described transformation steps, we created one separate MDA transformation project for each step. In order to work out the transformation rules it was very helpful to use examples given in the SBVR/XMI interchange syntax. For prototyping we used an example provided by Donald Baisley modeling a manager-employee relationship (about 400 line of XMI code) [12]. The vocabulary of Baisley's example is limited to about 20 percent of the SBVR vocabulary, including many frequently used fact types like `ConceptHasDesignation` or `FactTypeHasRole`. All in all we have mapped 39 SBVR fact types and 37 SBVR noun concepts. Despite the limitation the example is a good base for debugging and getting verifiable results. It will be necessary to use further examples to expand the transformation scope.

5   Summary

We presented a transformational technique to bridge the gap between SBVR based business rules and languages of the Semantic Web. Because all transformations between languages are completely based on the languages abstract syntax (metamodels) our approach is a contribution to model-driven integrity engineering. As target languages we chose OWL for the mapping of SBVR vocabulary and R2ML for the mapping of SBVR rules. We could show that metamodel based transformations using QVT are possible to implement with available tools in the Eclipse environment. R2ML seems to be an appropriate choice because there are already multiple ways to interpret and execute R2ML rules by rule engines [18] such as ILOG or Oracle BR (Fig. 8). So far, we only experimented with a subset of SBVR given by the Baisley's example. In the future, we have to find out which SBVR subset is appropriate for the automation of business rules. Furthermore the presented transformations have to be implemented by tools.

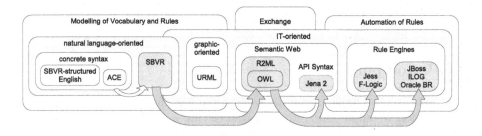

Fig. 8. Vision of automation of business rules

References

1. Object Management Group (OMG): Object Constraint Language Version 2.0 (formal/06-05-01), http://www.omg.org
2. Object Management Group (OMG): Production Rule representation (bmi/2007-03-05), http://www.omg.org
3. Object Management Group (OMG): Semantics of Business Vocabulary and Business Rules (SBVR) (dtc/06-08-05), http://www.omg.org
4. Wüst, K.: Business Rules auf Basis von UML/OCL. Diploma Thesis, Technische Universität Dresden (2006)
5. Spreeuwenberg, S., Gerrits, R.: Business Rules in the Semantic Web, Are There Any or Are They Different? In: Barahona, P., Bry, F., Franconi, E., Henze, N., Sattler, U. (eds.) Reasoning Web. LNCS, vol. 4126, pp. 152–163. Springer, Heidelberg (2006)
6. Object Management Group (OMG): MOF Query View Transformation (QVT) (ptc/05-11-01) (2005), http://www.omg.org
7. Wagner, G., Giurca, A., Lukichev, S.: A General Markup Framework for Integrity and Derivation Rules. Dagstuhl Seminar Proceedings 05371, Principles and Practices of Semantic Web Reasonning (2006)

8. WWW Consortium (W3C): RIF Use Cases and Requirements (July 2006), http://www.w3.org/TR/rif-ucr/

9. WWW Consortium (W3C): OWL Web Ontology Language Overview. (February 2004), http://www.w3.org/TR/2004/REC-owl-features-20040210

10. Object Management Group (OMG): Meta Object Facility (MOF) Core Specification (formal/06-01-01) Version 2.0 http://www.omg.org

11. Object Management Group (OMG): Ontology Definition Metamodel (ad/06-05-01) (2006), http://www.omg.org

12. Baisley, D.: A Small Taste of SBVR-based XMI. From email correspondence of OMG's SBVR Finalization Task Force (April 2006), http://www.omg.org/archives/sbvr-ftf/msg00300.html

13. The Eclipse Modeling Framework (EMF) Overview, http://dev.eclipse.org/

14. Halpin, T.A.: A Fact-oriented Approach to Schema Transformation. In: Thalheim, B., Gerhardt, H.-D., Demetrovics, J. (eds.) MFDBS 91. LNCS, vol. 495, Springer, Heidelberg (1991)

15. Reasoning on the Web with Rules and Semantics (REWERSE). Hompage (2007), http://www.rewerse.net

16. Liebau, H.-B.: Business Rules auf Basis des Semantic Web. Diploma Thesis. Technische Universität Dresden (2007)

17. EMF Ontology Definition Metamodel. IBM (September 2006), http://www.eclipse.org/

18. REWERSE I1 Rule Markup Language (R2ML), http://oxygen.informatik.tu-cottbus.de/rewerse-i1/

Take - A Rule Compiler for Derivation Rules

Jens Dietrich[1], Jochen Hiller[2], and Bastian Schenke[3]

[1] Massey University, Institute of Information Sciences and Technology,
Palmerston North, New Zealand
[2] Top-Logic Business Operation Systems GmbH, D-63263 Neu-Isenburg, Germany
[3] Brandenburgische Technische Universität Cottbus, Institut für Informatik, D-03013
Cottbus, Germany

Abstract. Rule engines have been used successfully in recent years in
order to improve the agility of enterprise applications. Most existing rule
engines focus on production rules, neglecting another important class of
rules, derivation rules. We point out that derivation rules are very useful
in many application scenarios, and present Take, a rule compiler for Java.
Take compiles derivation rules into optimized, reflective code that can
be deployed into running applications.

1 Introduction

The idea of considering rules as first class citizens in computing is not new. In
the areas of artificial intelligence and expert systems, complex rule based systems and entire programming languages were developed in the eighties. However,
the success of object oriented technology with its focus on modularization has
sidelined many of these approaches. In the last few years there has been a remark-
able comeback of rule based technologies with several vendors and open source
projects competing for market share, and several attempts to establish standards
for rule processing, exchange and storage [1,2,3]. This development is driven by
the need to shorten the software development life cycle that is perceived as too
slow to reflect the rapidly changing business environment. Externalizing rules
and putting them into data stores where they can be changed without entering another iteration of software development is a promising approach to make
systems more agile.

Most existing solutions are based on the idea of representing rule explicitly
as objects in a target programming language, and to process them using interpreters. These interpreters are also called inference engines. Inference engines use
different strategies for rule processing, the most popular being forward reasoning using the RETE algorithms [4]. This is suitable for production rules, that is,
rules where new assertions are computed if certain conditions are satisfied. This
means that an inference engine has to represent all rules and facts in memory
in order to propagate changes in the fact or rule base through the dependency
graph built from the rules.

Another rule category, derivation rules, has been widely neglected by the
business rule community. This is surprising as many business rules are derivation

A. Paschke and Y. Biletskiy (Eds.): RuleML 2007, LNCS 4824, pp. 134–148, 2007.

rules: they are mainly used to query instantiated domain models. Often they are hard coded in programming language specific functions or methods. Those often have names starting with prefixes such as *get*, *compute* or *calculate*. The classical example is the CRM system where the discount for a given customer has to be computed based on certain rules. This is a classical query, and not a state changing operation. It should also be noted that query driven rule systems fit very well into most business applications which are inherently pull based - processes are initiated by http requests and then use other query mechanisms (SQL queries, LDAP lookups) to compute results. State changing actions are then performed based on those results.

The second driver for rule based applications is the semantic web initiative of the W3C. The aim of this initiative is to turn the web into a large distributed knowledge base consisting of linked resources. Ontology languages like RDFS and OWL are used in order to describe resource types and their relationships. While these ontology languages have certain built-in rules, it has become apparent that there is a need to have custom defined derivation rules to reason about resources. A major use case is the use of derivation rules to establish the trustworthiness of resources. This is another query based use case: query a rule base to find out whether a given resource can be trusted or not. The semantic web initiative has triggered the development of a new generation of XML based rule markup languages, including RuleML [1], SWRL [5] and R2ML [6].

The rest of the paper is organized as follows: in the next chapter, we elaborate on some design goals for rule compilers. In chapter 3 we discuss the design of the Take compiler. Chapter 4 covers the integration of Take into applications. In chapter 5, various technical aspects of the actual compiler are discussed. The next chapter covers external fact stores, an important feature to ensure the scalability of Take. Finally we describe the integration of markup languages via the R2ML interface and finish with a brief discussion on related approaches.

2   Design Goals for Rule Compilers

2.1   Scalability

Enterprise applications may have a large number of rules. However, the real scalability issue is the even larger number of facts. Facts are associations between business objects. Typically, these facts are stored in relational databases from where applications fetch them as needed using (SQL) queries. Other sources include web services. It is important to emphasize that we do not argue that external facts should not be cached. This is often very useful and even necessary. However, the design of the rule compiler (and the same is true for inference engines as well) should not make it mandatory to cache the content of external fact stores. The storage problem is not the major problem here as prices for memory and hard disk space continue to decline. The main problem is the synchronization of cached data with the changing original data sources. This leads to the following design goals: facts should be referenced without the need to fetch them into memory, and the life cycle of connections to external fact sources must be managed. In particular, (database and network) connections have to be closed once they are not longer needed.

2.2 Reflectiveness

A major advantage of specifying application logic with rules is that the rules used to compute a certain result can be made available to the application. This is a major paradigm shift away from black box computing towards white box computing. Black box computing has been the dominating paradigm for many years, mainly driven by the desire to achieve a high level of modularization. For instance, encapsulation of objects is a core concept in object-oriented technology, and non-investigability through interfaces is part of a widely used definition of software components [7]. However, in many cases it would be very useful to ask a component why its method return certain values, such as "why does customer X get only $5 discount?" or "why is this email classified as junk mail?". If the application can return this information, further inquiries can be made if rules have meta data attached such as "who has entered this rule?" and "when has this rule been modified?". Finding this information in todays software requires detective skills to find the rules implemented in code and to trace its origin through design and requirement documents. Note that reflectivness is not about revealing implementation details of certain algorithms, but their abstract structure expressed in abstract, implementation independent terms. In this sense, reflectivness can co-exist with good OO practices such as encapsulation. Moreover, it can be seen as a natural extension of the reflection interface most programming languages offer. These interfaces provide runtime access to interfaces (inheritance trees, method signatures etc) whereas reflectiveness as introduced here provides runtime access to the semantics of artifacts.

2.3 Maintainability

One of the main problems with compiling rules into code is deployment. If applications have to be recompiled and redeployed to integrate code generated from rules, most of the advantage of using rules is lost. In particular, compiler based solutions face the issue of generating code and deploying it without interrupting the host application.

2.4 Separation of Specification and Implementation

This is a common principle of object-oriented design, facilitated by language features such as abstract classes and interfaces. This principle can be extended to rules: while rules may change often, the queries used by applications to consult the rules are relatively stable. This has certain consequences for the life cycle of applications using rules: applications may reference classes (or rather interfaces) generated for the query interfaces of rules, but not the classes representing the rules themselves. These classes must be dynamically loaded at system runtime.

3   Design

Take is a rule compiler for derivation rules that can generate Java artifacts from derivation rules, compile and deploy these classes. While the only output language supported at the moment is Java, Take is designed to support other target languages as well. Take consists of the following main components:

1. The knowledge base API. This is used to represent rules, facts and related concepts such as predicates, function and term. The API is instantiated by loading a knowledge base from a knowledge source such as Take scripts or XML files.
2. The Take scripting language. A simple to use scripting language that can be used to create and edit rules and therefore to instantiate the types in the API. It is integrated into Take as a knowledge source.
3. The R2ML interface. This is a module to import R2ML [6] instances into Take; it is also implemented as a knowledge source.
4. The compiler. This is essentially a Java source code generator that generates Java code from rules.
5. The knowledge base manager. Module that uses the compiler, compiles the generated classes into byte code and loads and instantiates them.

The overall design is similar to Java Server Pages (JSP). Similar to JPSs, with the life cycle of knowledge bases consisting of the following steps:

1. Rules are imported from an external source, parsed and the knowledge base API is instantiated.
2. The compiler generates Java source code from these objects (JSPs are translated into servlet sources).
3. The Java compiler is used to compile the generated sources into Java byte code.
4. These generated classes are loaded and instantiated.
5. Objects referenced in the knowledge base are bound.

This design is facilitated by the availability of several Java technologies. Class loaders can be used to load and unload classes dynamically; this supports the plug and play like deployment of components. The Java compiler API [8] can be used to compile classes at runtime. The loading of scripts is facilitated by the availability of parser generators and the Java scripting API [9]. All these features together make Java a good platform for dynamic, domain specific languages that can interact with application objects.

4   The Runtime Interface

Derivation rules are based on facts. Rules can be used to derive facts from other facts. The counterpart of facts in object oriented technologies are association

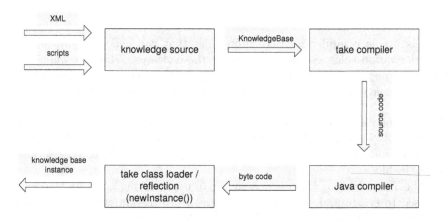

Fig. 1. Compilation and deployment knowledge bases

(for relationships between objects) and properties (for relationships between objects and primitive data values). Object-oriented languages like Java do not directly support associations, although it has been argued that associations should treated as first class citizens [10]. However, in UML, associations are explicitly represented. Associations are usually mapped to attributes (in the supplier type), with the type being either the client type (for one-to-one associations), or a container type (for one-to-many associations) that may use the client type as a generic parameter. Mapping associations to explicit association objects is also not uncommon, in particular if the associations have complex semantics [11,12]. This includes bidirectional associations that need synchronization, associations with attributes, and non-binary associations. The similarity between associations and predicates can be explored to use derivation rules in order to define the semantics of derived associations in class diagrams [13].

This motivates us in using association-like objects to represent facts. As far as queries are concerned, the generated classes representing the knowledge base use association classes in query results. For instance, consider the knowledge base defined in listing 1[1]. The knowledge base is define using the Take scripting language. This script defines two rules (in lines 9 and 11) and one query[2] (in line 8). Note that all lines starting with @ are syntactic sugar - they add meta data to queries, rules and the entire knowledge base that can be used by tools for different purposes. The informal meaning of the query is that there is a *discount* association between instances of *Customer* and *Discount* and that the application will provide a *Customer* instance but wants the *Discount* instance to be

---

[1] This example is in the Take subversion repository:
http://take.googlecode.com/svn/trunk/take/src/example/nz/org/take/compiler/
example1/crm-example.take, package names for classes are omitted for space reasons.

[2] For a complete specification of the scripting language, the reader is referred to the online documentation [14].

```
1  @@dc: creator=jens dietrich
2  @@dc: date=27/05/2007
3  var Customer customer
4  ref Discount goldCustomerDiscount
5  @take. compilerhint . class=CustomerDiscount
6  @take. compilerhint . slots=customer , discount
7  @take. compilerhint .method=getDiscount
8  query discount [ in , out ]
9  rule1 : if category [customer ," gold "] then discount [customer ,
10 goldCustomerDiscount]
11 rule2 : if 500<customer . turnover then category [customer ," gold " ]
```

Listing 1. Sample knowledge base defined using the Take scripting language

computed. That is, in the application the query results should be represented as instances of a class (lets say *CustomerDiscount*) representing the association between instances of *Customer* and *Discount*, respectively.

The [*in, out*] parameter in the query definition specifies the slots defined and the slots to be computed, the terms in the head of rule1 define the types of the object being associated. Objects are referenced as terms. As in first order predicate logic, there are three kind of terms:

1. Variables - placeholders for objects used in rules. Example: *customer* in listing 1.
2. Constants - constant objects represented by names that have to be bound to objects by the application. Example: *goldCustomerDiscount* in listing 1.
3. Complex Terms - terms that can be computed from other terms. Example: *customer.turnover* in listing 1.

Note that in general the cardinality of associations defined by rules in not known. That is, given a *Customer* instance c and a non-trivial set of rules it is neither known by the application programmer nor by the rule designer how many instances d of *Discount* exist such that $discount[c, d]$ holds. And even if there is exactly one solution, that is, one pair (c, d), there might still be different justifications for this result. The reason is that different rules have been applied, but the result is always the same in terms of equality of the objects involved. If we accept that these explanations themselves have meaning and value (see 2.2), then we have to regard pair computed using different rules as different. This suggests the use of a collection like query interface. On the other hand, many applications are not interested in all but only in some results. Moreover, computing all results might be computationally expensive. Therefore, the Take compiler uses an iterator (and not a collection) as the interface to query the associations computed by rules. Applications can fetch as many results as they need, with results computed using lazy initialization. If prefetching is required for performance reasons, the buffered iterator pattern [15] can be employed. Take defines two interfaces to represent results:

1. *nz.org.take.rt.ResourceIterator* extends *java.util.Iterator* and adds a single method *close*(). The semantics of *close*() is that it releases all resources allocated by the iterator, in particular connections to external fact stores (see section 6 for details).
2. *nz.org.take.rt.ResultSet* extends *nz.org.take.rt.ResourceIterator* and provides an API that can be used by applications in order to query the rules used to compute the current result. This API does not return the rules themselves but only the ids. Applications can use these ids as keys to retrieve rules and their meta data from the knowledge base.

For the example, the compiler will generate the following class representing query results.

```
 1 public class CustomerDiscount {
 2     public Customer customer;
 3     public Discount discount;
 4     public CustomerDiscount() {
 5         super();
 6     }
 7     public CustomerDiscount(
 8         Customer customer,
 9         Discount discount
10     ) {
11         super();
12         this.customer = customer;
13         this.discount = discount;
14     }
15 }
```

Listing 2. Class generated for predicate *discount* defined in listing 1

The compiler uses a naming class to compute the names for the Java artifacts to be generated. The default naming implementation uses annotations starting with the prefix @*take.compilerhint* (see listing 1) if present.

5   The Compiler

The actual compiler is defined by the interface *nz.org.take.compiler.Compiler*, the reference implementation is *nz.org.take.compiler.reference.DefaultComp-iler*. As discussed in 2.4, the compiler must support two functions. Firstly, the compiler must be able to generate interfaces and classes supporting those interfaces. Secondly, the compiler must support the generation of the actual knowledge base that can be used to handle queries.

The compiler has some properties that can be used to customise it. A naming class is used to generate the names of the artifacts to be generated, the location class is used to define the physical location of the source an compiled classes to be generated. The package and class name properties define the fully qualified name of the main class or interface to be generated.

5.1 Generating Interfaces

The compiler has a method $compileInterface(KnowledgeBase)$ that will generate the following artifacts from a knowledge base (that is loaded from a script or other source):

1. An interface for the knowledge base class to be generated.
2. For each predicate referenced in a query a class representing this predicate.

For instance, the following script generates the interface.

```
 1 Compiler compiler = new DefaultCompiler ();
 2 compiler.setNameGenerator(new DefaultNameGenerator());
 3 compiler.setLocation (new DefaultLocation ());
 4 InputStream script = // points to Take script file
 5 ScriptKnowledgeSource ksource =
 6   new ScriptKnowledgeSource (script);
 7 compiler.setPackageName (
 8   "example.nz.org.take.compiler.example1.spec");
 9 compiler.setClassName("DiscountPolicy");
10 compiler.compileInterface(ksource.getKnowledgeBase());
```

Listing 3. Script to generate the interface for the knowledge base defined in listing 1

The interface generated from the script in snippet 1 is listed in snippet 4 (package declaration omitted):

5.2 Generating Implementation Classes

The second function of the compiler is to generate implementation classes. While the compiler can be used to generate only the sources of the implementation classes, often another more convenient utility, the knowledge base manager ($nz.org.take.deployment.KnowledgeBaseManager$) is used instead. The

```
 1 public interface DiscountPolicy{
 2     public ResultSet <CustomerDiscount> getDiscount (
 3         Customer customer
 4     );
 5 }
```

Listing 4. Interface generated for the knowledge base defined in listing 1

knowledge base manager combines the compiler with the Java compiler and a custom class loader in order to automate the process depicted in figure 1. Using the knowledge base manager, the knowledge base interface that has been generated at system design time and is referenced by application classes can be instantiated dynamically with a few lines of code. At this point in time, objects referenced by name in the script have to be bound.

```
1  DiscountPolicy KB = null; // this is the generated interface
2  // compile and bind constants referenced in rules
3  KnowledgeBaseManager<DiscountPolicy> kbm =
4     new KnowledgeBaseManager<DiscountPolicy >();
5  Bindings bindings = new SimpleBindings ();
6  bindings.put(" goldCustomerDiscount" ,new Discount(20 ,true ));
7  InputStream script = // points to Take script file
8  KB = kbm.getKnowledgeBase(
9          DiscountPolicy.class ,
10         new ScriptKnowledgeSource(script ),
11         bindings );
```

Listing 5. Script to generate, load and instantiate the implementation class

Once the interface has been instantiated applications can use the simple iterator interface to fetch results as shown in the next listing.

```
1  Customer john = new Customer(" John ");
2  john.setTurnover(1000);
3  ResultSet <CustomerDiscount> result = KB.getDiscount(john );
4  System.out.println("The discount for John is: " +
5     result.next().discount );
```

Listing 6. Script to query the knowledge base

The generated code is fairly complex and details are outside the scope of this paper. The underlying algorithm used is resolution [16]. That is, there is an (implicit) agenda of goals that must be proven, and goals are matched to methods representing facts and rules using unification. The entire design is based on the Iterator design pattern [17], and more precisely on classes implementing *java.util.Iterator*. The main reason for this is that computation is triggered by the application requesting more results. This is done by invoking the *next*() and *hasNext*() method. Methods are generated for queries, with queries being defined as predicate/ parameter signature combinations. That is, a query consists of a predicate and a description which of its parameters are supplied by the application, and which parameters are to be computed. For each query, the following methods are generated:

1. A method that returns a chained iterator. The overall result for a certain query is an iterator built by concatenating the results obtained from performing the methods associated with rules and facts supporting the respective query.
2. For each rule that uses the query predicate in its head, a recursively nested iterator is built that represents the tree structure of the derivation tree. An inner class is generated to keep track of the variable bindings within the rule.
3. For each fact that uses the query predicate, a method is generated that returns either a singleton iterator (if the fact matches the query) or an empty iterator (otherwise).

A small runtime package contains some abstract iterator classes that are subclassed in the generated code. These classes are similar to some iterator implementations that are part of the Apache Commons Collections package [18]. The main difference is that they use generic parameters, and implement the extended *ResourceIterator* interface.

5.3   Backtracking and Logging Derivations

Another class from the runtime package, *nz.org.take.rt.DerivationController* is used for two purposes: managing backtracking and recording the derivation log. This class must be available at runtime so that applications can query the application log in order to address the requirements discussed in 2.2. The following code listing continues listing 6 and shows how the application can retrieve the rules and facts used in the derivation.

```
1  System.out.println("Rules  and  facts  used:");
2  for  (DerivationLogEntry  e:result.getDerivationLog()) {
3      System.out.println(e.getName());
4  }
```

Listing 7. Script to query the derivation log (follows listing 6)

5.4   Unification

Rules usually have unbound variables. These variables reflect the general character of the rules. However, the variables have to be replaced by objects when the results are computed. This is done as follows:

1. An inner class representing the bindings for all variables that occur in the rule and all complex terms containing variables is created and instantiated.
2. Constant terms are bound to static fields in a generated *Constants* class. These fields are initialized using the variable bindings provided by the application.
3. If the method generated for the rule is invoked, the parameters declared as input parameters are bound to the method parameters.

4. Whenever all variables that are part of a complex term have bindings, the value of this complex term is computed and the complex term is bound to the computed value.
5. When the methods associated with the prerequisites of the rule are performed, the remaining terms are successively bound.

Once all terms have bindings, the next result can be returned to the application. Note that the computation of complex terms is possible only because the functions used are themselves Java artifacts. Take supports two kind of functions:

1. Methods. Java methods that can be invoked.
2. Properties. Properties have getters (accessors) that can be invoked.

For instance, consider *rule2* in listing 1. This rule contains the complex term *customer.turnover*. If this rule is queried with a certain instance of *Customer*, lets say *aCustomer*, the binding *customer* → *aCustomer* will be added. In the next step, *customer.turnover* can be bound to *aCustomer.getTurnover()*. The reason is that *turnover* is recognized as Java by the knowledge sources and the generated code can reference the accessor of this method. Note that reflection is only used when the knowledge source (the script in this case) is imported, and not when the generated methods are invoked.

6   External Fact Stores

As pointed out in 2.1 it is necessary to reference external fact stores without explicitly fetching their complete content into memory. For this purpose, Take can defines external fact stores. Fact stores are managed in knowledge bases alongside rules and facts. They are identified by unique ids and have a single property defining a predicate. This is the predicate used in all facts provided by the fact store. In listing 8, a simple scenario is defined with a *grandfather* rule[3]. The basic *father* facts are imported from an external fact store.

```
1 var grandchild , father , grandfather
2 query is_grandfather_of [ out , in ]
3 rule1 : if is_father_of [ grandchild , father ]
4    and is_father_of [ father , grandfather ]
5    then is_grandfather_of [ grandchild , grandfather ]
6 external facts1 : is_father_of [ Person , Person ]
```

Listing 8. Script defining a query, a rule and a fact store

The compiler will generate an interface for the external fact store with methods returning resource iterators for the type generated for the fact store predicate. Applications have to implement this interface and pass it to the knowledge

---

[3] The full example can be found here: http://take.googlecode.com/svn/trunk/take/src/ test/nz/org/take/compiler/scenario8/

base manager. These implementations are essentially iterator factories. The iterators have to be resource iterators in order to allow the knowledge base calling back the *close()* method. This method can be used to close underlying database and network connections. The generated methods have parameters representing all predicate slots. If the knowledge base method invoking this method does not have bindings for these parameters, *null* will be used. This fact can be used in order to optimise the queries. For instance, in the example given above the fact store implementation can use the *father* parameter in an SQL query to fetch only the relevant records, avoiding a much more expensive "fetch all" query. However, if both parameters were *null* the "fetch all" would be necessary. In the scenario given, this would be the case the knowledge base was queried for all person-grandfather relationships. Note that Take does not index the facts - this is done by the external fact stores. Most facts stores, in particular relational databases, have highly optimised and mature techniques to do this.

```
1  public interface ExternalFactStore4IsFatherOf {
2      // Get all instances of this type from the fact store.
3      public nz.org.take.rt.ResourceIterator<IsFatherOf> fetch(
4          Person son,
5          Person father);
6  }
```

Listing 9. External fact store interface created by the compiler

7   The R2ML Interface

Take contains an interface to import knowledge from R2ML [6] sources[4]. R2ML is a new rule markup language that combines ideas from functional languages (such as OCL [19]) with relational rule languages (such as' SWRL [5]). It is designed as an interchange format for rules that supports different language paradigms to allow interchange of rules without loss of information. The R2ML working group provides web services for the translation between the more important rule languages and R2ML [20] including RuleML [1], SWRL [5], OCL [19], F-Logic XML [21] and Jena [22] rules. With the R2ML interface, Take can use these other rule languages as knowledge sources.

The import of R2ML sources into a Take knowledge bases is implemented in the class *nz.org.take.r2ml.R2MLDriver* that accepts a *Reader* instance pointing to the R2ML source as parameter, and controls the import process. The driver supports some customizations, for example to map data types and add slot names to predicates. Furthermore, there is a class *nz.org.take.r2ml.R2ML-KnowledgeSource* that works similar to the *ScriptKnowledgesource* discussed earlier.

---

[4] The R2ML adapter is a separate module in the project repository:
http://take.googlecode.com/svn/trunk/R2MLAdapter

The basic concepts from R2ML are similar to Take. Both systems support facts/atoms, constants/literals, predicates and variables. The vocabulary part of R2ML is not supported by Take. Object types in R2ML are mapped to Java classes (beans) that have to be provided by the application[5]. The mappings are defined with a *DataTypeMapper* that is passed to the *R2MLDriver*. Most of the features could be mapped directly from R2ML to take. R2ML *AttributionAtoms* are mapped to *PropertyPredicates* in Take. Property predicates are predicates backed by a Java property. R2ML *Functions* are mapped to *JFunctions*. *AssociationAtoms* don't have a direct counterpart in the Java/-Take world and are mapped to *SimplePredicate* in Take. Simple predicates are predicates defined by rules. That is, these predicates occur in rule heads.

The mapping from R2ML to Take is not trivial because of differences in expressiveness. R2ML rule bodies are much more complex than Take rule bodies. In particular, the R2ML adapters used by the R2ML driver accept only conditions in disjunctive normal form. To achieve this, the driver uses a normalizer that is responsible for the normalization of conditions. A simple default strategy is implemented that throws an exception in case the condition is not already provided in DNF. For each disjunct in the condition a single Take rule is created. Also, Take supports only one fact in the head of a rule. Therefore, R2ML rules with conjunctions in rule heads have to be split into several rules with atomic rule heads. There are some features in R2ML that can't be mapped to a single fact (prerequisite) but only to a list of facts. Examples are *ObjectDescriptionAtoms* and *AssociationAtoms*). In general, R2ML is a more compact format and the mapping results in greater number of rules in Take. However, for most applications these rules are de-facto invisible as they are part of an intermediate model between rule markup and compiled code.

8   Conclusion, Related Approaches, and Future Work

We have presented Take, a rule compiler that compiles rules into executable Java code and has the ability to deploy this code into running systems. We related core design decisions to the four general design goals of scalability, reflectivness, maintainability and separation of specification and implementation, and outlined its interface to rule markup languages. These features makes Take a good choice for numerous applications where fast deductive reasoning is required. As derivation rules can be used to model problems on an abstract level, code generation tools for models fit well into MDA [23] scenarios where executables are generated from models. For instance, Take can be used as code generator for URML [13] models.

Take contains a number of interesting features not discussed here for space reasons, including its support for negation and built-in verification. The interested reader is referred to the project home page [24].

---

[5] If these classes do not exist, simple code generators can be used to generate these classes from the vocabulary.

There are many projects and products that try to integrate rules with object-oriented technologies. As far as we are aware of, most of them are based on inference engines utilising RETE [4] or similar algorithms. One vendor, Corticon[6] claims to have a product that compiles rules using a Design-Time Inferencing engine (DeTI^{TM}). We were not able to obtain more information about this.

Take is inspired by Mandarax [25], a framework for derivation rules in Java that uses a Prolog style inference engine. Much of the theory about linking objects and formal logic has been pioneered by F-Logic [26].

There are a number of features still missing in Take, in particular negation support for predicates solely defined by derivation rules, and built-in aggregation functions[7]. These issues will be addressed in future versions. There is also ongoing work on implementing the complete UServ business rule model [27] that is used for benchmarking rule engines. This will allow us to better compare the expressiveness and scalability of Take with other products. There are no planes to implement the JSR94 [2] specification. JSR94 is a suitable API for rule engines based on production rules. However, such a runtime API is less useful for a compiler based approach.

Acknowledgments

The authors would like to thank Jevon Wright and Jonathan Giles for their support.

References

1. Boley, H., Tabet, S., Wagner, G.: Design Rationale for RuleML: A Markup Language for Semantic Web Rules. In: SWWS, pp. 381–401 (2001)
2. JSR 94: (Java^{TM}Rule Engine API), http://www.jcp.org/en/jsr/detail?id=94
3. OMG: Business Semantics of Business Rules Request For Proposal (2003)OMG Document: br/2003-06-03
4. Forgy, C.L.: Rete: A fast algorithm for the many pattern/many object pattern match problem. Artificial Intelligence 19(1), 17–37 (1982)
5. Horrocks, I., Patel-Schneider, P.F., Boley, H., Tabet, S., Grosof, B., Dean, M.: SWRL: A semantic web rule language combining OWL and RuleML. Technical report, W3C Member submission 21 may 2004 (2004)
6. Wagner, G., Giurca, A., Lukichev, S.: A general markup framework for integrity and derivation rules. In: Bry, F., Fages, F., Marchiori, M., Ohlbach, H.-J. (eds.)

---

[6] http://www.corticon.com/

[7] Aggregation functions are needed to express rules like the following (from [27]):

1. ES_01a: If car is Not Eligible then increase policy eligibility score by 100.
2. ES_01b: If car is Provisional then increase policy eligibility score by 50.
3. ES_04: If eligibility score is less that 100, then client is eligible for insurance.

In the last rule, the results of querying the first two rules have to be aggregated.

Principles and Practices of Semantic Web Reasoning. Dagstuhl Seminar Proceedings, IBFI Schloss Dagstuhl, Germany. Dagstuhl Seminar Proceedings, vol. 05371 (2006)

7. Szyperski, C.: Component software: beyond object-oriented programming. ACM Press/Addison-Wesley Publishing Co, New York (1998)
8. JSR 199: (Java™ Compiler API) http://jcp.org/en/jsr/detail?id=199
9. JSR 223: (Scripting for the Java™ Platform), http://jcp.org/en/jsr/detail?id=223
10. Rumbaugh, J.: Relations as semantic constructs in an object-oriented language. In: OOPSLA 1987. Conference proceedings on Object-oriented programming systems, languages and applications, Orlando, Florida, United States, pp. 466–481. ACM Press, New York, NY, USA (1987)
11. Noble, J.: Basic relationship patterns. In: EuroPLOP 97: Conference proceedings(1997)
12. Genova, G., del Castillo, C.R., Llorens, J.: Mapping UML Associations into Java Code. Journal of Object Technology 2(5), 135–162 (2003)
13. Rewerse Working Group I1 - Rule Modeling and Markup (URML - a UML-Based Rule Modeling Language),http://oxygen.informatik.tu-cottbus.de/rewerse-i1/?q=node/7
14. Dietrich, J.: The Take manual (2007), http://docs.google.com/Doc?id=dgzzp6gn_23fx23xx
15. Brase, R.: Eliminate bottlenecks with a buffered iterator (2002), http://builder.com.com/5100-6370-1046648.html
16. Robinson, J.A.: A machine-oriented logic based on the resolution principle. J. ACM 12(1), 23–41 (1965)
17. Gamma, E., Helm, R., Johnson, R., Vlissides, J.: Design patterns: elements of reusable object-oriented software. Addison-Wesley Longman Publishing Co., Inc., Boston, MA, USA (1995)
18. Apache Foundation (Apache Commons Collections), http://jakarta.apache.org/commons/collections/
19. OMG: Object Constraint Language Specification, version 2.0. Object Management Group (2006)
20. Rewerse Working Group I1 - Rule Modeling and Markup (Web Service for Rule Interchange), http://oxygen.informatik.tu-cottbus.de/rewerse-i1/?q=node/27
21. de Bruijn, J., Kifer, M.: D16.2v0.1. F-logic/XML - An XML Syntax for F-logic (2004), http://www.wsmo.org/2004/d16/d16.2/v0.1/20040324/
22. Carroll, J., Dickinson, I., Dollin, C., Reynolds, D., Seaborne, A., Wilkinson, K.: Jena: Implementing the semantic web recommendations (2003)
23. OMG (OMG Model Driven Architecture), http://www.omg.org/mda/
24. Take (project home page), http://code.google.com/p/take/
25. Dietrich, J.: A rule-based system for ecommerce applications. In: KES, pp. 455–463 (2004)
26. Kifer, M., Lausen, G., Wu, J.: Logical foundations of object-oriented and frame-based languages. Journal of the ACM 42(4), 741–843 (1995)
27. Business Rule Forum: (UServ Product Derby Case Study), http://www.businessrulesforum.com/2005_Product_Derby.pdf

The OO jDREW Engine of Rule Responder: Naf Hornlog RuleML Query Answering

Benjamin Larry Craig

Faculty of Computer Science, University of New Brunswick
Frederiction New Brunswick, Canada
`ben.craig AT unb.ca`

Abstract. Rule Responder is an intelligent multi-agent system for collaborative teams and virtual communities that uses RuleML as its rule interchange format. The system allows these virtual organizations to collaborate in an effective semi-automatic manner, and is implemented as a Web-based application on top of the Enterprise Service Bus Mule. It supports rule execution environments (rule/inference engines) such as the Prova distributed Semantic Web rule engine and OO jDREW (Object Oriented java Deductive Reasoning Engine for the Web). The paper describes the role of OO jDREW for answering queries against rule bases in the Naf Hornlog RuleML sublanguage for the real-world use case of a symposium organization.

1 Introduction

Person-centered and organization-centered profile descriptions using Semantic Web techniques such as Resource Description Framework (RDF), Friend of a Friend (FOAF)[1], Semantically-Interlinked Online Communities (SIOC) and ExpertFinder has become increasingly popular. Recent work on FindXpRT [BLB+] generalized fact-based to rule-based profiles to capture conditional person-centered metadata such as the right phone number to call a person depending on the time, the topic, or the urgency. Rule Responder is a service-oriented middleware tool that can be used by virtual organizations for automated rule-based collaboration [PKB07]. Distributed users (human or automated agents) can interact with Rule Responder by simple query-answer conversations or complex negotiation and coordination protocols. The Rule Responder agents will process the events, queries, and requests according to their rule-based decision and behavioral logic, delegate certain subtasks to other agents, collect partial answers, and send the completed answer(s) back to the requestor. The communication protocol used between the architectural components of Rule Responder (e.g., external, personal, and organizational agents) is Reaction RuleML [PKB07]. The Rule Responder team is focused on implementing use cases that require the interchange of rule sets and the need for a querying service. The use cases demonstrate rule-based collaboration in a virtual organization. One specific use case

---

[1] http://www.foaf-project.org/

A. Paschke and Y. Biletskiy (Eds.): RuleML 2007, LNCS 4824, pp. 149–154, 2007.
© Springer-Verlag Berlin Heidelberg 2007

that will be explained in detail throughout the paper is the organization of the RuleML-2007 Symposium. Beside the contribution of an intelligent autonomous agent layer on top of the current Semantic Web, the use case demonstrates rule interchange between rule inference services using a common rule exchange format (RuleML/Reaction RuleML).

Section 2 of the paper explains the details of the RuleML-2007 use case. Section 3 of the paper contains details on the architectural structure of Rule Responder. In Section 4, an explanation of how Rule Responder and OO jDREW [BC] function is demonstrated. Section 5 will conclude the paper.

2   RuleML-2007 Use Case Description

The initial use case created to demonstrate Rule Responder is the organization of a symposium such as the RuleML-2007 Symposium, which is an example of a virtual organization that requires online collaboration within a team. Rule Responder can support the Organizing Committee of the RuleML-2007 Symposium by embodying responsibility assignment, automating first-level contacts for information regarding the symposium, and screening of incoming submissions based on metadata (e.g., to see if the paper topics fit the symposium or not). It can also aid with other issues associated with the organization of a symposium, including presentation scheduling, room allocation, and special event planning.

The use case utilizes a single organizational agent to handle the filtering and delegation of incoming queries. Each committee chair has a personal agent that acts in a rule-constrained manner on behalf of the committee member. Each agent manages personal information, such as a FOAF profile containing a layer of personal information about the committee member as well as FOAF-extending rules. These rules allow the personal agent to automatically respond to requests concerning the RuleML-2007 Symposium. Task responsibility in the organization is managed through a responsibility matrix, which defines the tasks committee members are responsible for. The matrix and the roles assigned within the virtual organization are defined by an OWL Lite Ontology. The Pellet [HPSM] reasoner is used to infer subclasses and properties from the ontology.

3   Architecture and Workflow of Rule Responder

Rule Responder's architecture is implemented by the use of personal agents, organizational agents, communication middleware and external agents. The rule-based personal agents represent different members of the RuleML-2007 Organizing Committee. These agents are implemented by a rule engine, which acts as an inference and execution environment for the rule-based decision and behavioral logic of the semi-autonomous agents. The organizational agents constitute an intelligent filtering and dispatching system, using a rule engine execution environment, for blocking incoming queries or delegating them to other agents. The communication middleware implements an Enterprise Service Bus (ESB) supporting various transmission protocols (e.g., JMS, HTTP, SOAP).

The ESB implementation for Rule Responder is Mule [BCC+], an efficient open source communication middleware. External agents can interact with the Rule Responder-enabled virtual organization via its public communication interface (e.g., an HTTP endpoint interface of an organizational agent as single point of entry).

The **personal agents** used by Rule Responder contain FOAF-extending profiles for each person of the organizational team. Beyond FOAF-like facts, person-centric rules are used. All clauses (facts and rules) are serialized in Naf Hornlog RuleML [HBG+], the RuleML sublanguage for Horn logic (allowing complex terms) enriched by Naf (Negation as failure). These FOAF-extending profiles have access to RDF (BibTeX, vCard, iCard, Dublin Core) and RDFS/OWL (role and responsibility models). The following is a query that a personal agent can be expected to answer: "What is the best (fastest response time) method to contact a committee member?" For answering it, the corresponding personal agent's FOAF profile must be queried to find the most convenient contact method.

The **organizational agent** contains a rule set that describes the organization. The following is a query that an organizational agent can be expected to answer: "Who should be contacted about the symposium's panel discussion?" When this incoming query is received by the organizational agent, the agent must first determine who the correct contact person is for the panel discussion. When the contact person has been confirmed, the organizational agent sends a subquery to that committee member's personal agent. The personal agent could then respond with a contact method (e.g., email or telephone number, depending on contact preferences in their FOAF profile). Alternatively, if that contact person was on vacation or currently busy, then the personal agent would respond back to the organizational agent that the contact person is unavailable. If the first-line contact person cannot be reached, then the organizational agent will use the responsibility matrix (i.e., which committee members are responsible for certain tasks, and what members can fill their role of if they are unavailable) to try to contact the next personal agent.

External agents can communicate with the virtual organization by sending messages that transport queries, answers, or complete rule sets to the public interface of the organizational agent (e.g., an HTTP port to which **post** and **get** requests can be sent from a Web form). The standard protocol for intra-transport of Reaction RuleML messages between Rule Responder agents is JMS. HTTP SOAP is used for communication with external agents, such as Web services or HTTP clients (Web browsers).

4   OO jDREW Execution for the RuleML-2007 Use Case

The internal agents of Rule Responder are currently implemented using the reasoning engines Prova [KPS] and OO jDREW[BBH+05]; although further rule engines can be adapted for use by Rule Responder (which basically requires a translator between RuleML and the execution syntax of the rule engine). OO jDREW is used for certain personal agents, while Prova is used for other personal agents and for the organizational agent. OO jDREW has two main modes of operation: top-down and bottom-up. Bottom-up execution is used to infer all derivable knowledge from a set of clauses (forward reasoning). Top-down execution is used to solve a query on the knowledge base (backward reasoning). Rule Responder primarily uses top-down execution due to the nature of query-answering services.

The query-answering service begins when a message from an external agent is received by an organizational agent. To obtain a query from the exchanged message, the OO jDREW engine must first parse the message and translate the message payload (i.e. the RuleML query) into the executable syntax of OO jDREW. OO jDREW then loads that person's FOAF profile (stored as POSL [Bol04] or RuleML), and according to the rules defined therein, derives the answer, which is sent back to the requester as a Reaction RuleML event message. Once the FOAF profile has been loaded, an RDF Schema file of user classes will be required if the FOAF profile requires a user defined taxonomy. When the FOAF profile and RDF Schema has been parsed and the query was successfully parsed, then OO jDREW will run the query against the knowledge base. After solutions have been derived, they must be serialized into a Reaction RuleML message, to be sent back to the requesting agent. The following is an example of such a query-answering service being executed by Rule Responder. As mentioned in section 3, we want to know: "Who should be contacted about the symposium's panel discussion?" When the person to be contacted is determined a subquery is used to get the contact information of that person. The messages below show the interchange between the agents.

```
Incoming Message to Organizational Agent from the Rule Responder External Agent (shown left):
Incoming Message (subquery) to OO jDREW from the Organizational Agent (shown right):
<RuleML>
  <Message mode="outbound" directive="query">                    . . .
    <oid><Ind>RuleML-2007</Ind></oid>
    <protocol><Ind>esb</Ind></protocol>
    <sender><Ind>Ben Craig</Ind></sender>                 <content>
    <content>                                               <Atom>
      <Atom>                                                  <Rel>person</Rel>
        <Rel>getContact</Rel>                                 <Var>Name</Var>
        <Ind>ruleml2007_Panel</Ind>                           <Var>Role</Var
        <Ind>update</Ind>                                     <Var>Title</Var>
        <Var>Contact</Var>                                    <Var>EMail</Var>
      </Atom>                                                 <Var>Telephone</Var>
    </content>                                              </Atom>
  </Message>                                              </content>
</RuleML>                                                      . . .
```

The subquery corresponds to the head of a rule (in POSL), executed in OO jDREW, whose body premises could be directly retrieved as facts or further delegated to other agents. This is just a sample rule, for a full FOAF profile please refer to this reference [Cra].

```
person(?Person, ?Role, ?Title, ?Email, ?Telephone) :-
  contact(?Person, ?Email, ?Telephone),
  role(?Person, ?Role),
  title(?Person, ?Title).
```

Each message contains a performative wrapper that is used by Rule Responder to understand how to interpret and use the message. The information contained in the message is comprised of the following: the *sender* of the message, the transport *protocol* used by the ESB, the object identifier (*oid*), the *content* of the message, and the attributes *mode* and *directive*. The content of the incoming messages are the queries that are to be answered by either a personal or organizational agent. The content of the outgoing messages are the query results. The *oid* is the conversation id of the current message, and is used to distinguish between multiple conversations. The *mode* attribute indicates whether the message is an inbound or outbound message. The *directive* attribute distinguishes whether the message's content is a query or an answer.

When this message (above, left) is received by the RuleML-2007 organizational agent, the agent can determine which Organizing Committee members are to be contacted about the panel discussion. The agent will then send a subquery (above, right) to the personal agent (OO jDREW servlet). Once the query has been successfully executed, OO jDREW will send the following message (below, left) back to the RuleML-2007 organizational agent.

```
Outgoing content from the OO jDREW Servlet to the Organizational Agent (shown left):
Outgoing content from Organizational Agent to the External Agent (shown right):
    <content>                              <content>
      <Atom>                                 <Atom>
        <Rel>person</Rel>                      <Rel>getContact</Rel>
        <Ind>John</Ind>                        <Ind>ruleml2007_Challenge</Ind>
        <Ind>PanelChair</Ind>                  <Ind>update</Ind>
        <Ind>PHD</Ind>                         <Expr>
        <Ind>John@email.com</Ind>                <Fun>person</Fun>
        <Ind>1-555-555-555</Ind>                 <Ind>John</Ind>
      </Atom>                                    <Ind>PanelChair</Ind>
    </content>                                   <Ind>PHD</Ind>
                                                 <Ind>John@email.com</Ind>
                                                 <Ind>1-555-555-555</Ind>
                                               </Expr>
                                             </Atom>
                                           </content>
```

The organizational agent will then parse the subquery solution and send one final message (above, right) back to the external agent. As can be seen in the example, the organizational agent starts a subconversation with the personal agent by asking a new query and uses the received results to answer the original query of the external agent. As the answer to a query, the query will be instantiated, i.e. each variable within the query will be bound to the found individual constant or complex term.

5   Conclusion

Rule Responder can be used to implement a wide range of use cases that require an intelligent, semi-automated decision layer (e.g. to collect data, infer new knowledge, answer queries, and solve problems). The middleware of Rule Responder allows deployment of multiple running use cases concurrently. The ESB provides the communication backbone to synchronously or asynchronously interchange messages between multiple agents. RuleML is a descriptive rule interchange language that can implement all logical structures that are necessary for Rule Responder. For more information about Rule Responder and a use case demo, readers are referred to [PBKC] (section "Use Cases"). Rule Responder is an open source project and its current rule engines, Prova [KPS] and OO jDREW [BC], are available open source as well.

References

[BBH+05] Ball, M., Boley, H., Hirtle, D., Mei, J., Spencer, B.: The OO jDREW reference implementation of ruleML. In: Adi, A., Stoutenburg, S., Tabet, S. (eds.) RuleML 2005. LNCS, vol. 3791, pp. 218–223. Springer, Heidelberg (2005)

[BC] Ball, M., Craig, B.: Object Oriented java Deductive Reasoning Engine for the Web. http://www.jdrew.org/oojdrew/

[BCC+] Borg, A., Carlson, T., Cassar, A., Cookeand, A., Fenech, S., et al.: Mule. http://mule.codehaus.org/display/MULE/Home

[BLB+] Boley, H., Li, J., Bhavsar, V.C., Hirtle, D., Mei, J.: FindXpRT: Find an eXpert via Rules and Taxonomies. http://www.ruleml.org/usecases/foaf/findxprt

[Bol04] Boley, H.: POSL: An Integrated Positional-Slotted Language for Semantic Web Knowledge (May 2004), http://www.ruleml.org/submission/ruleml-shortation.html

[Cra] Craig, B.: A RuleML-2007 Publicity Chair FOAF Profile. http://www.jdrew.org/oojdrew/rulesets/RuleML-2007publicityChair.posl

[HBG+] Hirtle, D., Boley, H., Grosof, B., Kifer, M., Sintek, M., Tabet, S., Wagner, G.: Naf Hornlog XSD. http://www.ruleml.org/0.91/xsd/nafhornlog.xsd

[HPSM] Hendler, J., Parsia, B., Sirin, E., et al.: Pellet: The Open Source OWL DL Reasoner. http://pellet.owldl.com/

[KPS] Kozlenkov, A., Paschke, A., Schroeder, M.: Prova: A Language for Rule Based Java Scripting, Information Integration, and Agent Programming. http://www.prova.ws/

[PBKC] Paschke, A., Boley, H., Kozlenkov, A., Craig, B.: Rule Responder: A RuleML-Based Pragmatic Agent Web for Collaborative Teams and Virtual Organizations. http://www.responder.ruleml.org

[PKB07] Paschke, A., Kozlenkov, A., Boley, H.: A Homogenous Reaction Rule Language for Complex Event Processing. In: EDA-PS 2007. Proc. 2nd International Workshop on Event Drive Architecture and Event Processing Systems, Vienna, Austria (September 2007)

Querying the Semantic Web with SWRL

Martin O'Connor, Samson Tu, Csongor Nyulas, Amar Das, and Mark Musen

Stanford Medical Informatics, Stanford University School of Medicine,
Stanford, CA 94305
martin.oconnor@stanford.edu

Abstract. The SWRLTab is a development environment for working with
SWRL rules in Protégé-OWL. It supports the editing and execution of SWRL
rules. It also provides mechanisms to allow interoperation with a variety of rule
engines and the incorporation of user-defined libraries of methods that can be
used in rules. Several built-in libraries are provided, include collections of
mathematical, string, and temporal operators, in addition to operators than can
be used to effectively turn SWRL into a query language. This language
provides a simple but powerful means of extracting information from OWL
ontologies. Used in association with a relational data importation tool that we
have developed called DataMaster, this query language can be also used to
express knowledge-level queries on data imported from relational databases.

1 Introduction

The Semantic Web project is a shared research plan that aims to provide explicit
semantic meaning to data and knowledge on the World Wide Web [1]. Semantic Web
applications aim to be able to integrate data and knowledge automatically through the
use of standardized languages that describes the content of Web-accessible resources.
OWL was developed as a formal language for constructing ontologies that provide
high-level descriptions of these web resources [2]. Recent work has concentrated on
adding rules to OWL to provide an additional layer of expressivity. The Semantic
Web Rule Language (SWRL; [3]) is one of the results of these activities. SWRL
allows users to write Horn-like rules that can be expressed in terms of OWL concepts
and that can reason about OWL individuals.

We have been developing open-source tools to work with SWRL. One of the
primary results is the SWRLTab [4, 9], an extension to the Protégé-OWL ontology
development toolkit [5]. Recently we have been concentrating on developing
mechanisms to allow SWRL to be used as an OWL query language. We are using this
language to query both standard OWL ontologies and ontologies that are populated
from relational databases. We have developed a tool called DataMaster [6] to perform
this importation. We believe these tools can use used to tackle some of the
challenging data integration issues that are faced when developing Semantic Web
based software.

A. Paschke and Y. Biletskiy (Eds.): RuleML 2007, LNCS 4824, pp. 155–159, 2007.

2 SWRLTab

The SWRLTab [4] provides a set of APIs that support the building of tools that work with SWRL rules. It has several software components, including (1) an editor that supports interactive creating, editing, reading, and writing of SWRL rules; (2) a rule engine bridge that provides the infrastructure necessary to interoperate with third-party rule engines and reasoners; (3) a built-in bridge that provides a mechanism for defining Java implementations of SWRL built-ins; and (4) a set of built-in libraries containing mathematical, string, temporal, and ABox and TBox operators.

2.1 SWRL Editor

The Protégé-OWL SWRL Editor is an extension to Protégé-OWL that permits interactive editing of SWRL rules. Users can create, edit, and read/write SWRL rules. With the exception of arbitrary OWL expressions, this editor supports the full set of language features outlined in the SWRL Submission. It is tightly integrated with Protégé-OWL and is primarily accessible through a tab within it. When editing rules, users can directly refer to OWL classes, properties, and individuals within an OWL ontology. They also have direct access to a full set of built-ins described in the SWRL built-in specification and to all of the XML Schema data types.

2.2 Rule Engine Bridge

The SWRL Rule Engine Bridge is a subcomponent of the SWRLTab that provides a bridge between an OWL model with SWRL rules and a third party rule engine or reasoner. Its goal is to provide the infrastructure necessary to incorporate rule engines and reasoners into Protégé-OWL to execute SWRL rules.

The bridge provides mechanisms to (1) import SWRL rules and OWL classes, individuals, properties and descriptions from an OWL ontology; (2) write that knowledge to a rule engine or reasoner; (3) allow the rule engine to perform inference and to assert its new knowledge back to the bridge; and (4) insert that asserted knowledge into an OWL ontology. The bridge also provides mechanisms to add graphical user interfaces to the SWRLTab to allow interaction between a particular rule engine implementation and users. A bridge to the Jess rule engine [7] is provided together with a user interface component called the SWRLJessTab.

2.3 Built-in Bridge

SWRL provides a very powerful extension mechanism that allows the use of user-defined methods in rules [8]. These methods are called built-ins and are predicates that accept one or more arguments. Built-ins are analogous to functions in production rule systems. A number of core built-ins are defined in the SWRL specification. This core set includes basic mathematical operators and built-ins for string and date manipulations. SWRL users can also define their own built-in libraries. Example libraries could include built-ins for currency conversion, or for statistical, temporal or spatial operations. Again, once implemented, these user-defined built-ins can be used directly in SWRL rules.

We have developed an extension to the SWRLTab called the SWRL Built-in Bridge. This extension provides support for defining built-in implementations written in Java and dynamically loading them. Users wishing to provide implementations for a library of built-in methods can define a Java class that contains definitions for all the built-ins in their library. The bridge has a dynamic loading mechanism to import these built-in definitions and provides an invocation mechanism to execute these loaded definitions from rule engines. A mechanism to marshall and unmarshall arguments to and from built-in is also provided. If additional rule engines are integrated into the SWRLTab they can use these existing built-in libraries without modifying them.

2.4 Built-in Libraries

Using the built-in bridge we have implemented a set of libraries for common methods required by rules [10]. These include implementations for the core SWRL built-ins defined by the SWRL submission, a temporal library that can be used to reason with temporal information in SWRL rules, and libraries with ontology TBox and ABox operators.

3 Extending SWRL to Support OWL Queries

We have also developed a built-in library that allows SWRL rules to be used to query OWL ontologies [11]. The library contains SQL-influenced built-ins that can be used in a rule to construct retrieval specifications. For example, the following rule, written with these built-ins, retrieves all persons in an ontology whose age is less than 5, together with their ages:

```
Person(?p) ^ hasAge(?p,?a) ^ swrlb:lessThan(?a,5) → query:select(?p,?a)
```

This query will return pairs of persons and ages. The following query lists all persons together with their ICD9 codes:

```
Person(?p) ^ hasICD9(?p, ?icd) → query:select(?p, ?icd)
```

This query will return pairs of persons and their ICD9 codes. Assuming a person can have more than one ICD9 code, multiple pairs would be displayed for each person—one pair for each code.

The query library also provides basic counting, aggregation, ordering, and duplicate elimination operators.

Query built-ins can be used with other built-in libraries provided by the SWRLTab. For example, the TBox built-in library can be used in SWRL queries to extract the structure of an OWL ontology; the temporal built-in library can be use to express complex temporal conditions in queries. The ability to use built-ins freely in a query provides a means of continuously expanding the power of the query language.

3.1 SWRLQueryTab

The SWRLQueryTab provides a convenient way to visualize the results of these SWRL queries. It has a control sub-tab that can be used to control the execution of SWRL rules containing query built-ins. A query can be selected from the rule table in the Protege-OWL SWRL Editor and executed to display results of the query. Users can navigate to that sub-tab to review the results displayed in tabular form.

3.2 SWRLQueryAPI

The SWRLQueryAPI provides a JDBC-like Java interface to retrieve the result of SWRL queries. Results for a particular query can be retrieved from a SWRL rule engine bridge. Rows in a result can be iterated through and the contents of each column in a row retrieved using accessor methods.

For example, if we wish to process the results of the earlier query (which we will name "Query-1") that extracts all persons under the age of 5 from our ontology, we can write:

```
Result result = bridge.getQueryResult("Query-1");

while (result.hasNext()) {
  System.out.println("Person: " + result.getDatatypeValue("?p"));
  System.out.println("Age: " + result.getDatatypeValue("?age"));
  result.next();
} // while
```

4 DataMaster

Importing data from relational databases into ontologies is frequently required, particularly when an ontology is used to semantically describe the data used by a software application. Another growing category of applications requires database-ontology integration and/or interoperation, where a mapping between the database schema structure and ontology concepts is the main focus. In the latter cases the import of the data residing in relational databases may not be necessary or desired.

To meet these requirements, we have developed DataMaster [6], a Protégé-OWL plug-in that allows the user to import relational database structure or content into an OWL ontology. DataMaster can be used with any relational database with JDBC/ODBC drivers. A user interface is provided that supports user-driven configuration of the importation process. A user can use this interface to connect to a database and select the portions of the database that they wish to import. If the user decides on a schema-only import the database schema is read and represented in OWL using the Relational-OWL [12] ontology. If a content import is requested, the use can select the a number of mapping options, such as, for example, how table and column names are mapped, and how relational column types are mapped to XSD Schema types.

5 Conclusion

We are using these tools to help meet the data specification requirements of several ontology-driven biomedical applications. SWRL is used to help unify the domain-level specification of system data with the run-time operational data needs of system components. In conjunction with OWL, SWRL is provides both a formal domain-level description of data in the biomedical systems that we are developing and a run-time query mechanism to execute domain-level queries on data in these systems. We have also implemented a dynamic OWL-to-relational mapping mechanism that allows data to be selectively retrieved from live operational database in response to SWRL queries [13]. These extensions will be release in an upcoming Protégé -OWL release. The ability to meet the high data throughput demands of these applications is crucial to enable this technology to meet the scalability requirements of the Semantic Web.

Acknowledgements. This work was supported in part by the Immune Tolerance Network, which is funded by the National Institutes of Health under Grant NO1-AI-15416, and also by the Centers for Disease Control and Prevention under grant number SPO-34603. We thank Valerie Natale for her editorial comments.

References

1. Berners-Lee, T.: The Semantic Web, Scientic American (May 2001)
2. OWL Overview: http://www.w3.org/TR/owl-features/
3. SWRL Submission: http://www.w3.org/Submission/SWRL/
4. SWRLTab: http://protege.cim3.net/cgi-bin/wiki.pl?SWRLTab
5. Knublauch, H., Fergerson, R.W., Noy, N.F., Musen, M.A.: The Protégé OWL Plugin: An Open Development Environment for Semantic Web applications. In: Third International Semantic Web Conference, Hiroshima, Japan (2004)
6. DataMaster: http://protegewiki.stanford.edu/index.php/DataMaster
7. Jess: http://herzberg.ca.sandia.gov/jess/
8. SWRL Built-in Specification: http://www.daml.org/rules/proposal/builtins.html
9. O'Connor, M.J., Knublauch, H., Tu, S.W., Grossof, B., Dean, M., Grosso, W.E., Musen, M.A.: Supporting Rule System Interoperability on the Semantic Web with SWRL. In: Fourth International Semantic Web Conference, Galway, Ireland (2005)
10. SWRL library: http://protege.cim3.net/cgi-bin/wiki.pl?SWRLTabBuiltInLibraries
11. SWRL queries: http://protege.cim3.net/cgi-bin/wiki.pl?SWRLQueryBuiltIns
12. de Laborda, C.P., Conrad, S.: RelationalOWL - A Data and Schema Representation Format Based on OWL. Conceptual Modelling 43, 89–96 (2005)
13. O'Connor, M.J., Shankar, R.D., Tu, S.W., Nyulas, C., Parrish, D.B., Musen, M.A., Das, A.K.: Using Semantic Web Technologies for Knowledge-Driven Querying of Biomedical Data. In: AIME 2007. 11th Conference on Artificial Intelligence in Medicine, Amsterdam, Netherlands (2007)

Implementation of Production Rules for a RIF Dialect: A MISMO Proof-of-Concept for Loan Rates

Tracy Bost[1], Phillipe Bonnard[2], and Mark Proctor[3]

[1] Forum Technologies, dba Valocity, 7714 Poplar STE 200, Germantown, Tennessee, USA
tbost@valocity.com
[2] ILOG SA, 9, rue de Verdun BP 85, 94253 Gentilly Cedex
pbonnard@ilog.fr
[3] Red Hat UK Ltd, 200 Fowler Avenue, Farnborough, Hampshire, GU 14 7JP, UK
mproctor@redhat.com

Abstract. In June 2006, the Mortgage Industry Maintenance Organization (MISMO) began an initiative to help facilitate the electronic exchange of business rules between trading partners in the mortgage vertical. The Business Rules Exchange Workgroup (BREW) was subsequently formed to generate a charter and lead the adoption within the industry. BREW set out to create a proof-of concept (POC) to solve an often mentioned need for the exchange of rules: loan application pricing. A modified Production Rules (PR) extension to the working draft of the Rules Interchange Format (RIF) core design was developed by ILog and the well established MISMO schema was used as the ontology to share and execute a ruleset among ILOG JRules & JBOSS Rules in a distributed environment. A ruleset consisting of pricing rules were created using a decision table. ILOG Business Rules Engine(BRE) serialized these rules into a RIF ruleset. Using web services over HTTP , this ruleset was made available for consumption via web services allowing a client tool to consume the ruleset and push to a JBOSS Rules engine that had been exposed as a web service. Once the two systems shared the same ruleset, data payloads could then be sent to each rules engine for rule execution.

Keywords: Business Rules Exchange, MISMO, Production Rules Interchange, Rules Interchange Format, Rules Vendor Interoperability.

1 Introduction

The Mortgage Industry Standards Maintenance Organization (MISMO) is a subsidiary of the Mortgage Banker's Association and serves as the mortgage industry's body for electronic standards. In June of 2006, MISMO created the Business Rules Exchange Workgroup (BREW) to help create a standard in the industry for the interchange of business rules among trading partners.

To demonstrate to the industry the possibility of exchanging business rules is in the realm of today's technical capabilities, the group decided to conduct a proof-of concept (POC). In order for the POC to be meaningful, it was decided that it should be performed using an existing documented use case scenario. Additionally, it was

A. Paschke and Y. Biletskiy (Eds.): RuleML 2007, LNCS 4824, pp. 160–165, 2007.

desired the POC would consist of at least two different vendors and platforms to add to the credibility (a reality) of the task.

Several potential use cases were discussed. However, it was the" pricing rules" use case that seemed to be the one most often mentioned as a candidate. The pricing rules were thought to be simple enough to be expressed with an existing rule dialect, while being able to utilize the MISMO vocabulary without resorting to ontologies outside the domain of MISMO.

The W3C Rules Interchange Format (RIF) core design working draft was chosen as the base dialect to be used in the proof-of concept. This decision was based on the fact some participates in the MISMO BREW workgroup were active RIF working group as well. Additionally, RIF had recently made publicly available its first working draft.

In its most common denominator, pricing is the interest rate a borrower will pay on the mortgage loan. Due to the complexity of a mortgage loan, several factors could affect this "pricing". Probable attributes to determine a borrower's pricing would be credit rating, loan to value percent (LTV), whether the borrower has full documents(proof of income) or simply stated (no proof of income) documentation, and the type of loan (conventional, refinance, etc.) sought by the borrower.

A lender or broker will typically obtain pricing rules from a document downloaded from various investor or lender web sites, by electronic mail, or through facsimile. Once the pricing rules are in hand, someone manually inputs the rules into a business rules management system (BRMS). Obviously, this can be a time consuming as well as a potentially error prone process.

A more efficient approach would be to have a standard, semantic way for an investor or lender to share these pricing rules with its trading partners. It is possible this could save the industry millions of dollars in cost savings and from loss of business.

The scope of the proof-of concept was confined to interchange and execution of a subset of business rules typically used in a real world scenario. Since current pricing rules are manually sent from a publisher (investor or lender) to various subscribers (lenders or brokers), it was decided a publish-subscribe model in a distributed environment would suffice the needs of the project. When each Business Rules Engines (BRE) located on separate servers obtained the same ruleset, a MISMO compliant data payload would be sent to each BRE via web services for execution.

2 Technical Implementation

ILOG agreed to develop the core extension and interfaces for the translation drivers and titled the library APOCRIF, which is an acronym for "A Proof-of-Concept Rules Interchange Format". The packages in this library were developed using Java 5.0. ILOG and JBOSS Rules both shared the same Java model and processed data supplied by the MISMO version 2.4 schema.

2.1 APOCRIF Overview

APOCRIF explores the possibility of serializing and de serializing an ILOG rules language ruleset as an XML RIF document. The only common data model used by the

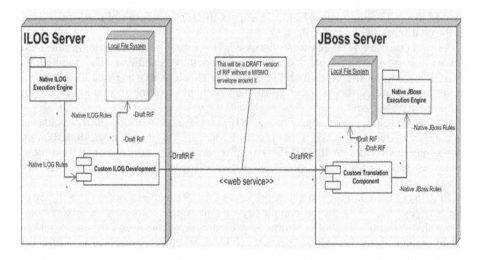

Fig. 1. This is a high level architecture diagram for the MISMO Proof-of Concept. [1]

rule engines was the provided MISMO schema. The RIF core design working draft is limited to Horn clauses [3]. As such, to complete a successful integration, it was necessary to introduce some small tweaks to the RIF core design, along with minor adjustments to the MISMO schema. As one may expect, the RIF meta-model had to be extended as well.

2.2 Global Architecture

The translation process was centered on the meta-model of RIF. This meta-model was represented by a set of classes defined in the **apocrif.core** Java package. Those factory classes help represent a RIF ruleset as a Java object tree.

2.3 Rule Engine Drivers and Extensions for Production Rules

On top of the **apocrif.pr** package, a driver layer was introduced. This layer charged the adapting of the RIF rules and their meta-model to the vendor's rule language.

This driver layer additionally provided for an interface to handle the serialization and de serialization of the RIF ruleset.

The **apocrif.pr** package is where the extensions to the RIF core design for Production Rules were introduced as well. The ProductionRule class was created as a subclass of the core Rule class. The Rule class was extended by adding Java Beans for "IF" & "THEN" parts, priorities, and constraints. Binary terms and XML getter & setters were additional extensions included in the **apocrif.pr** package.

2.4 Rule Mapping

A minimiumistic approach was used in the rule mapping as to only create the necessary extension modules to the RIF core design. ILOG created a rule mapping table relating the "Production Rule" concept to the appropriate "RIF" concept.

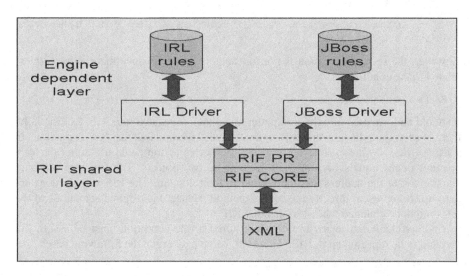

Fig. 2. This diagram shows how the meta-model provides an important XML-RIF serializati on/de serialization service that can be utilized by different rules engines

The MismoRoot type is the mapping type of an XML complex type defined in the MISMO schema. It was assumed that every type appearing in a rule expression is related to an XML type from that schema.

2.5 The Rules Interchange Client Tool

The APOCRIF User Interface client tool was developed with Java 5.0 . The tool needed to provide good visuals as the target audience was from a business background. A canvas with a drag-n-drop functionality seemed like a good choice for that reason. To help save time for users of the tool, a pre-defined "rules interchange job" (RIJ) was created prior to the demonstration. The RIJ was simply an XML based file containing the community (BREs in this case) connections, and components that are part of the rule interchange project. Loaded onto RIJ Designer canvas were XML Payloads, Rule Agents, and the APOCRIF controller components.

3 Analysis

This proof-of concept has given those interested to have a glimpse of what challenges may lay ahead. For example, the RIF version used for APOCRIF (http:// www.w3.org/2005/rules/wg/core/draft-20070323.html) requires that an expression be related to a variable. For example, consider the first rules condition expression:

```
x: A(x==y);
```

This is translatable into RIF as something similar to:

```
Forall (declare: Var (name: x) Var (name: y),
```

```
    formula: Implies ( if: Equals( side: Var( name:x )
Var (name:y) ) )   )
```

However, the second condition is not translatable into RIF since there is no variable related to the condition A.

```
A (this==y);
```

It should be formatted into the first pattern, and then translated into RIF. Additionally, there is no way to disquinish between the "equals" method and the "= = "operator. In a JRules rule, a value expression of a condition or an action could use either the '= =' operator or the equals Java method to compare two objects. The first one refers to comparison of the address and the second to their content. The RIF version does not permit it to represent directly the equals method. Hence, the original semantics of the rules might be changed when serialized in RIF.

Another challenge seems to be it is required a rule condition must be assigned a variable to be represented in RIF. However, some may argue the following case:

```
fn:compare(comp1,comp2)
```

```
fn:compare(comp1,comp2,collation)
```

This would Return -1 if comp1 is less than comp2, 0 if comp1 is equal to comp2, or 1 if comp1 is greater than comp2 Arguing the same case using numerical values could be represented as the following:

`op:numeric-equal`	`Equality comparison`
`op:numeric-less-than`	`Less-than comparison`
`op:numeric-greater-than`	`Greater-than comparison`

A resulting conclusion could be the functions back up the "eq", "ne", "le" and "ge" operators in XPath. However, that is to say that a "le" operator located in a value expression should be translated into the XPath operator as a conjunction of the op: numeric-equal and an op: numeric-less-than operators, and not as a unique and fictive operator op: numeric-less-equal-than. It also creates complexities for the translation of a RIF expression into a PR expression since one should reduce the conjunction of two operators as one in order to get back the original expression.

Additional issues encountered from the implementation included[2]:

- Meta properties do not exist in RIF.
- There is no ruleset variable in RIF.
- The way the MISMO Schemas introduces union type for numeric values makes it difficult to support by Object Model Binding. These union types had to be replaced with xsd: int or xsd: double.
- The RIF core design provides no way to distinguish from a variable declaration and a variable value.
- As a result of staying with the RIF core design for elements to represent a production rules implementation, one can end up with very generic XML with no ability for semantic validation via an XML schema.
- There is difficulty navigating nested structures and the weakness of type such as not being able to differentiate decimal from currency.

4 Future Direction

Despite the challenges, the proof-of concept served its purpose well as it facilitated the successful exchange and execution of a ruleset from different BREs in a distributed environment. However, APOCRIF was a very early adaptation of a draft of the RIF core design. As a proof-of concept project, the team decided it would not spend too much time extending something that would be changed later. It is preferred that it represents most of the production rule instructions by an already existing and powerful RIF instruction, the Uniterm. As such, once these instructions for production rule exist, the POC will provide for rules modification and round tripping among the BREs as its next step.

Acknowledgments. The proof-of concept would not have been possible without the additional hard work of volunteers within the MISMO & Business Rules Engines communities. A special thanks to Doug Doedens, Chair of the MISMO Business Rules Exchange Workgroup, Mark Proctor, Redhat /JBoss (Drools) , and the ILOG team of Bruno Trimouille, Christian de Sainte Marie (co-chair WC3 Rules Interchange Format Working Group), and Michael Chen.

References

[1] Doedens, D., Bonnard, P.: MISMO WIKI, "Proof of Concept" (May 2007), http:// wiki. mismo.org/MISMOWiki/Wiki.jsp?page=ProofOfConcept
[2] Proctor, M.: W3C Rules Interchange Format for Production Rule Systems (June 01, 2007), http://markproctor.blogspot.com/2007/06/w3c-rule-interchange-format-for.html
[3] World Wide Web Consortium Rules Interchange Format Core Design Working Draft (March 2007), http://www.w3.org/2005/rules/wg/core/draft-20070323.html

Related Project Links:

Mark Proctor's Blog of the MISMO POC
http://markproctor.blogspot.com/search/label/MISMO
MISMO Wiki (Proof of Concept for Rules Exchange)
http://wiki.mismo.org/MISMOWiki/Wiki.jsp?page=ProofOfConcept
APOCRIF Data & Project Source Code
http://anonsvn.labs.jboss.com/labs/jbossrules/contrib/apocrif/

Adapting the Rete-Algorithm to Evaluate F-Logic Rules

Florian Schmedding, Nour Sawas, and Georg Lausen

University of Freiburg, Institute for Computer Science
Georges-Köhler-Allee 51, 79106 Freiburg, Germany
schmeddi@informatik.uni-freiburg.de
n.sawas@tu-bs.de
lausen@informatik.uni-freiburg.de

Abstract. The evaluation of production rules is typically based on the Rete-algorithm. The topic of the current paper is to investigate, whether set-oriented bottom-up evaluation of rules in deductive databases can also take advantage of the Rete approach. We report on our implementation of the Rete algorithm as one possible evaluation technique inside the F-Logic rule evaluation engine Florid. We demonstrate, that in situations in which several rules share common subgoals a considerable improvement of the execution time can be gained by the Rete approach. We show this by means of benchmark programs, also comparing our results with the performance of Jess, a production rule system relying on Rete.

1   Introduction

The Rete algorithm [3] is often used to implement production rule systems because it's known to scale well in scenarios with a lot of rules and few facts. In deductive databases, we observe the opposite: A vast amount of facts and comparatively few rules.

But does this mean that Rete is useless in deductive databases? Maybe it could be changed in a way to cope with them, too. As far as we know there aren't any reports on this question in literature. Therefore we implemented Rete in our F-logic reasoner Florid to investigate its performance on datalog programs. Our anticipation was that the evaluation time is reduced when rules share common parts in their bodies.

We compare our Rete-prototype to the semi-naive bottom-up evaluation, the standard evaluation approach in Florid. We don't care about the also available naive strategy because of its slowness. To get an idea of the quality of our Rete-implementation, we compare our speed with the state-of-the-art production rule system Jess.

The possible benefit of the Rete-evaluation would be the automatic runtime-optimization of programs, where we have many rules with common subgoals, maybe due to a less versed user, (semi-) automatic generation of rules or in

A. Paschke and Y. Biletskiy (Eds.): RuleML 2007, LNCS 4824, pp. 166–173, 2007.

cases when materialization of the common parts is not intended so that the rules cannot be rewritten.

Before we describe our implementation in detail, we give a brief overview about Rete and F-logic/Florid. We will finally demonstrate our benchmark results and show a considerable advantage of the Rete approach in our prototype.

2 F-Logic and Florid

Frame-logic (F-logic) [5,7] is a declarative language combining the advantages of deductive databases with object orientation (e. g. object identity, class hierarchy, inheritance, polymorphism). Although it has a first-order semantics, higher-order features like sets and inheritance can be expressed, too. Along with the objects (function symbols in first-order logic) also predicates are allowed, hence F-logic covers datalog. We implemented F-logic in our system called Florid (*F-logic reasoning in databases*) [6]. Path expressions were added to F-logic in [4].

Florid is a native implementation, i. e., it stores the objects (frames) directly in its object manager instead of converting everything to predicate logic. Florid evaluates bottom-up according to the fixpoint semantics, in the naive or seminaive way. To support negated cycles in the program, the user can divide it into several strata. The Florid program is also able to retrieve web pages. The page content gets stored in a DOM-like structure and can be used like ordinary F-logic facts.

3 The Rete-Algorithm

The Rete-algorithm [3] is an efficient algorithm to match patterns to an object base, mostly used in production rule systems, designed by Ch. Forgy. "Rete" is a Latin word for "net". The idea behind the algorithm is to transform all the rules of a given program into a network in order to reduce the programs' execution time by sharing common structures in the network. We outline this idea with a simplifying example.

```
p(a,b).    q(b,f).
p(c,b).    q(e,g).
p(b,e).
r(X,Y) :- p(X,Y), q(Y,U).
s(X,Z) :- p(X,Y), p(Y,Z), q(Z,U).
```

Figure 1 illustrates the different evaluation approaches for the given program. On the left side both rules are handled separately, therefore the join results (a, b), (c, b) and (b, e) are calculated twice, whereas the simplified rete-evaluation on the right hand side uses these results as the output of the first rule and as interim result for the second one.

The network consists of an alpha and a beta part. The former employs *constant test nodes* and *alpha memories* to compute matches for each literal of the rule set. The latter performs join operations and computes matches for rules within its *join nodes*, *production nodes* and *beta memories*.

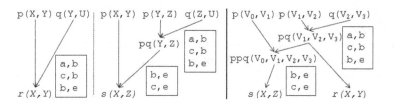

Fig. 1. Rule evaluation: individually (left) and based on Rete (simplified)

4   Implementation and Adaptation to F-Logic

For the implementation [9] of the Rete-algorithm we followed the basic description in [2]. But because of the broad field of applications, there seems to be no definitive Rete-implementation. For example, production rules allow the deletion of working memory elements (facts). The sequence of insertions and deletions is not known in advance, so the use of indexes is not always faster. There are also different enhancements of Rete available like Treat [8], Leaps [1] or Rete* [10]. As a general goal, we try to implement Rete in a set-oriented style for the database scenario.

4.1   WME and Working Memory

As we have mentioned above, we restricted out prototype to the datalog subset of F-logic. In addition, we also allowed the use of objects and functional methods without arguments. WMEs are arrays whose size depends on the corresponding predicate or method. For methods, e.g. $o[m \rightarrow r]$, the WME is (o, m, r), thus always having length three. With predicates, the size equals the predicates' arity plus one to store the predicate name (on the second position, just like the method name[1]). A predicate $p(x_1, x_2, \ldots, x_n)$ has the form $(x_1, p, x_2, \ldots, x_n))$ as WME.

The working memory is split into two hash tables, one for P-atoms and one for F-atoms. Their keys are the first fields of the WMEs (thus method or predicate name). Each key points to another hash table, whose keys are a hash function of the WME fields except the first field.

4.2   Token

Tokens represent a sequence of WMEs that represent a match to (a part of) a production – a sequence of i WMEs is a match for i conditions of a production. To save space, we implement tokens via pointers to the corresponding WMEs and already constructed tokens: A token of length $i + 1$ points to a WME that was joined with a token of length i, and a pointer to this token. So firstly we avoid to copy WMEs and secondly to store tokens of length i with each prolongation of a match as a copy in the new token with length $i + 1$.

---

[1] This actually allows variables over predicate names but that is prevented by the F-logic parser.

4.3   Memory Structures

Alpha memories are simple lists of all WMEs that passed the constant test nodes. This is a disadvantage because this list has to be searched for matches when a *left activation* (a new token arrives through a beta memory) occurs in a join node. We therefore implemented a hash table in the join nodes (see below).

The beta memories carry information about partial production matches in their tokens and should also be indexed on the position which will be tested in the next join node. When a beta memory has more than one child (join nodes), the different joins need different indexes. In this case we have to create a new beta memory and copy the tokens from the first beta memory to its hash table. This omits to re-compute the join operation of its predecessor.

As a difference to beta memories production nodes don't use indexes on their tokens (complete matches for a rule), but have additional information about the productions' head. That's the reason why they cannot be shared like beta memories. But to omit unnecessary computations of joins, we copy the contents of the production node to a new beta memory if another rule can re-use it.

4.4   Join Nodes

Join nodes are responsible for the combination of tokens to new matches. When a new token arrives in a beta memory, the attached children join nodes are left activated and search for matches in the related alpha memory. For this case we need an alpha memory index, which is stored in the join node. We opted for this because storing the index in the alpha memory would imply to have different copies of the same contents if different indexes are used. This in turn would require changes inside the constant test node design. On a right activation (a WME arrives through the alpha memory), matches are searched in the index of the corresponding beta memory.

As a difference to [2], we don't propagate the join results directly to the join nodes' successors. We are not interested in the immediate effect of a new WME but we want to compute all possible results of a given program. So we can evaluate set-orientedly, that is to say all possible new matches are computed before activating any successor.

5   Benchmark

To test our prototype, we developed a few programs and various datasets. We compared naive, semi-naive bottom-up and rete evaluation. Because of the poor performance of the naive approach we omit the results here. We ran the tests on a Windows XP PC with an Intel Pentium D 2.8 GHz CPU (Florid uses only one core) and 2 GB main memory. We used Jess 7.0RC1 with Java 1.6 for the Jess programs.

The semi-naive evaluation is applied twice: Primarily with the normal program, thereafter with stratification after every rule in order to check only one rule for possible new matches at a given time.

5.1 Datasets

For our extensional database (EDB) facts we generated a directed graph with 199 nodes and 176 edges. The nodes have abbreviations of country names as labels, the edges are borders between them. The input predicates have the form *border("de","pl")*, for instance. We extended this graph by multiplying the original one (renaming the nodes) and connecting the topmost node with the lowest of two different graphs in a binary-tree-style. This was done to have a certain connectivity in the graph and not a completely random one, but we didn't put attention to the possible regularities in it. A sketch is shown in Fig. 2.

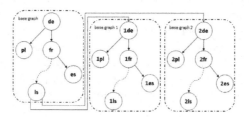

Fig. 2. Test data set

There are nine datasets of increasing size to test for an extra slowdown in the Rete-prototype compared to the semi-naive evaluation if the EDB increases. Table 1 presents the number of facts contained in each graph and the number of results deduced by the programs.

Table 1. The different graph sizes

Graph	# facts	# results
1	10088	338319
2	20000	671323
3	30089	1010056
4	40001	1343060
5	50090	1682228
6	60002	2015232
7	70091	2353965
8	80003	2686969
9	90092	3026137

5.2 Programs

The programs in our test set derive new facts by connecting more and more nodes. To test the effectiveness of rete, we used two different programs: in one program we optimized the rules by hand to re-use previous results, in the other program we did not (Rete should do it for us). We refer to these as *short* and *long* version. We'll only present the long program here, it's obvious from the context which predicates should be re-used. We didn't use recursion here (our prototype supports it).

```
parentOf(X,Y) :- border(X,Y).
childOf(X,Y) :- border(Y,X).
grandParentOf(X,Y) :- border(X,Z), border(Z,Y).
grandChildOf(X,Y) :- border(Y,Z), border(Z,X).
grandGrandParentOf(X,Y) :- border(X,Z), border(Z,U), border(U,Y).
grandGrandChildOf(X,Y) :- border(Y,Z), border(Z,U), border(U,X).
grandGrandGrandParentOf(X,Y) :- border(X,Z), border(Z,U),
                                border(U,V), border(V,Y).
grandGrandGrandChildOf(X,Y) :- border(Y,Z), border(Z,U),
                                border(U,V), border(V,X).
sibling(X,Y) :- border(Z,X), border(Z,Y).
spouse(X,Y) :- border(X,Z), border(Y,Z).
uncleOf(X,Y) :- border(Z,X), border(Z,U), border(U,Y).
greatUncleOf(X,Y) :- border(W,U), border(W,V), border(U,Z),
                     border(Z,Y), border(V,X).
niceOf(X,Y) :- border(Z,Y), border(Z,W), border(W,X).
cousin(X,Y) :- border(V,X), border(U,V), border(U,W), border(W,Y).
```

Datasets and programs were automatically converted to the Clips-syntax of Jess. We also had a version using the allowed object-syntax, but because of the very different implementation in the rete prototype and Florid we found it not to be very comparable. Jess doesn't support objects either.

5.3   Results

All basic results are shown in Tab. 2 (time is given in seconds). Pure semi-naive evaluation has clearly the worst performance. The Rete version has an advantage over the stratified program, which is also much faster than the semi-naive one. Although not quite as fast, Jess also seems to have a good performance. The difference may be due to Java in contrast to C++, but at least the VM heap size was big enough to avoid garbage collection as much as possible.

Table 2. Benchmark-results: **(a)** short, **(b)** long program version

short	semi-naive	stratified	rete	jess
1	8,1	4,3	3,3	8,0
2	21,4	9,1	6,8	16,0
3	44,0	14,0	10,3	24,0
4	61,1	18,9	13,7	31,0
5	89,9	23,8	17,2	42,0
6	138,1	29,4	21,0	49,0
7	187,1	34,6	24,3	56,0
8	197,7	40,5	28,0	63,0
9	249,4	45,7	31,7	75,0

long	semi-naive	stratified	rete	jess
1	9,4	6,7	4,8	13,0
2	26,0	14,0	9,6	27,0
3	60,3	21,1	14,6	40,0
4	89,5	29,4	19,8	54,0
5	163,3	38,4	24,7	67,0
6	235,0	46,2	29,9	84,0
7	249,4	54,0	35,4	97,0
8	332,6	64,1	40,6	110,0
9	526,5	76,1	46,4	123,0

The absolute time values are not too significant owing to the different implementations: Only the semi-naive and the stratified programs use the same internal structures. Though being part of Florid, the Rete-prototype uses the data structures adapted from [2] to store facts and not the internal object manager of Florid. Needless to say, Jess uses another programming language.

To do better, we take the time increment of each evaluation strategy to qualify them. By the means of Fig. 3, we can assume a linear increase of the evaluation time except for the semi-naive version. Our prototype has the smallest increase.

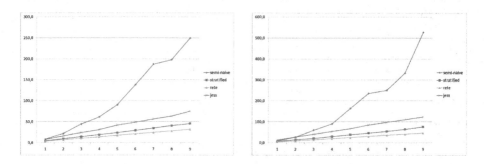

Fig. 3. Time increase of **(a)** short and **(b)** long versions

6   Conclusions and Future Work

In this paper we described the application of the Rete-evaluation algorithm in the field of deductive databases. We implemented it in our F-logic reasoner Florid, where we concentrated mainly on the datalog subset of this language.

We implemented slight modifications to the basic algorithm to improve its performance, for example, the re-use of production nodes and hash tables for alpha memories. We also compute all results of a join at once before activating its successors to have a more set-oriented evaluation.

As our timing results show, our prototype had the fastest evaluation for the test programs. There was no performance drawback with increasing dataset sizes. This shows that there are cases in which re-using join results with Rete is an efficient way to improve performance.

Compared to Jess, our prototype was faster, but the time differences between the short and long program version were bigger in Jess. It seems that Jess exploited more redundancies than our prototype. At the moment we search only from left to right to find common structures, that's the reason why not all rules can be optimized in the above test program. In programs with rules like

```
grandParentOf(X,Y) :- border(X,Z), border(Z,Y).
grandParentOf2(X,Y) :- border(X,Z), border(Z,Y).
```

and

```
grandParentOf(X,Y) :- border(X,Z), border(Z,Y).
grandParentOf2(X,Y) :- border2(X,Z), border2(Z,Y).,
```

the second rule in the upper case being a copy of the first rule, our prototype and Jess have a similar timing results. However, the upper program takes almost as long as the lower one instead of taking only about half the time. Beside the implementation of a better strategy to find redundancies, this is another point which has to be examined further.

The big differences in evaluation time between the semi-naive and the stratified programs also demonstrate the need for an efficient method to compute the conflict set. In the stratified version we have to look at only one rule per time, therefore it runs faster than the semi-naive program. A disadvantage is that a rule is lost after stratification and cannot be used anymore. This is not the case for Rete, nevertheless its conflict set computation seems to be fast.

There are still some tests to be carried out, especially with recursive programs, but the results up to now confirm the utility of Rete in deductive databases. We are now working on a tighter integration of Florid and the Rete-prototype to support full F-logic with Rete. Apart from that we seek to avoid Retes' materializing of intermediate results in such a way that they are removed from memory as soon as no rule depends on them any longer.

Of course, the test programs used here contain many redundancies which make Rete look good. We took them to get an idea of our implementation but it is certainly not the case for many other applications. We don't believe that Rete is a better general solution, instead we seek to integrate it with the seminaive evaluation to apply it whenever redundancies are found.

References

1. Batory, D.: The LEAPS Algorithms.Technical Report 94-28, Department of Computer Sciences, University of Texas at Austin (1994)
2. Doorenbos, R.: Production Matching for Large Learning Systems. PhD thesis, Computer Science Department, CMU, Pittsburgh (1995)
3. Forgy, C.L.: Rete: a fast algorithm for the many pattern/many object pattern match problem. Artificial Intelligence 19, 17–37 (1982)
4. Frohn, J., Lausen, G., Uphoff, H.: Access to objects by path expressions and rules. In: VLDB 1994. Proceedings of the 20th International Conference on Very Large Data Bases, pp. 273–284. Morgan Kaufmann, San Francisco, CA, USA (1994)
5. Kifer, M., Lausen, G., Wu, J.: Logical foundations of object-oriented and frame-based languages. J. ACM 42(4), 741–843 (1995)
6. May, W.: Florid User Manual. Technical report, Institut für Informatik, Universität Freiburg (2000)
7. May, W.: How to write F-Logic Programs in Florid. Technical report, Institut für Informatik, Universität Freiburg (2000)
8. Miranker, D.P.: TREAT: a new and efficient match algorithm for AI production systems. Morgan Kaufmann, San Francisco (1990)
9. Sawas, N.: Adapting the Rete-Algorithm to F-logic and its Implementation in Florid. Institut für Informationssysteme, Technische Universität Braunschweig, master's thesis (2007)
10. Wright, I., Marshall, J.: The execution kernel of RC++: RETE*, a faster RETE with TREAT as a special case. International Journal of Intelligent Games and Simulation 2(1), 36–48 (2003)

Rule Definition for Managing Ontology Development

David A. Ostrowski

System Analytics and Environmental Sciences
Research and Advanced Engineering
Ford Motor Company
dostrows@ford.com

Abstract. This paper presents an approach to ontology development through the application of declarative logic programming. Our method employs rules for the purpose of prototyping new ontology versions by decoupling the process of concept definition from the application of descriptive logics (DL) and advanced class representations. By generating new ontology versions on-the-fly we can test updates to the ontology design. This employment of rules expands on current efforts of translation and merging of ontologies. By employing this technique, we can support a pragmatic approach to the management and integration of instance data thus realizing a rapid-prototyping approach to the testing of potential updates to ontologies. Examples of this technique are presented utilizing a subset of the OWL-DL specification through the implementation of the Jena API. Advantages include the rapid testing of updated ontology representations (including the efficient remapping of instance data) and an efficient means of Ontology querying. Eventual benefits include Ontology versioning support and tool development to support the automatic engineering of instance data.

Keywords: OWL-DL, Description Logics, Declarative Logic Programming, Rules, Ontology development.

1 Introduction

In support of Semantic Web applications, there has been increased interest in the development of ontologies layered on top of legacy applications. The development of these ontologies represents a substantial challenge regarding integration with conventional software projects [1]. The intent is to utilize rules to infer updated representations to support the reconstruction or modifications of an earlier ontology version. By implementing a rule-based approach within a dynamic programming environment we allow developers to explore possibilities of ontology prototypes without immediately addressing validation constraints that exist in current development environments. By expanding the validation process to a series of steps, we can achieve levels of efficiencies in cases where a large amount of information is stored within instance (ABOX) relationships. This work employs techniques of ontology translation and merging at a lower level (single ontology) than suggested by earlier work [2][3]. Through our layering of rules we intend to further break down the

A. Paschke and Y. Biletskiy (Eds.): RuleML 2007, LNCS 4824, pp. 174–181, 2007.

steps of ontology definition and refinement. These layers can also serve in the capacity of a self-documenting effort between ontology versions as well as support a stronger debugging activity.

The rest of the paper is organized as follows. Chapter two presents background information as motivating principles for this work. Chapter three presents real-world examples demonstrating use-cases of our rule-based approach. Chapter four presents a conclusion and description of future work.

2 Background

Ontology development can be divided into a series of steps [4][5]. This process begins with the basic identification of concepts (objects). Once the objects have been identified then relationships are defined beginning with the description of the domain and range. These definitions lead into complete applications of description logic formalisms. From this point we can define the instance data relationships. By breaking down this level of refinement we are interested in providing a higher level of efficiency to this development process.

2.1 Rule Applications

Ontology development and rule languages have long been identified for many activities within the Semantic Web. Rules have been applied in a number of specific areas including language layering, querying, data integration and Semantic Web search [6]. Work in the past demonstrating the value of merging ontologies and rule based systems have drawn comparisons between the intersections of rules and ontologies. Rules can be defined within a number of different categories. Among these characterizations of rules includes deductive inferences (ie. Application of basic if-then based formalisms in rule based constructs), transformations (where schemas/ontologies can be mapped between each other) constraint rules (supporting premises that A and B can't both be true) and condition/action rules (where if you see pattern A then do action B) [7].

2.2 DL and DLP

With some overlap in their functionality, a number of comparisons have been drawn between the use of DL and rule-based implementations further strengthening their complementary nature [8]. Description logics represent a highly constrained language (defined as a W3C recommendation) while rule-based systems represent a higher level of expressiveness [9][10]. As a result DL is easier to maintain while rule-based systems can result in complicated implementations, especially in cases of poor standard enforcement. Rule-based technology has been developed over 20 years ranging across many formalized implementations within the AI–based community. Here, the rule-based technologies provide strong enterprise support through the employment of robust technologies thus complementing recently developed DL based implementations.

Early work has identified mutual benefits of DL in the area of data management. Specifically, this has included enhanced access to data and knowledge by using descriptions in languages for schema design, integration, queries, updates, rules and constraints [11]. This work started with the proposal of simulation of descriptive logics within the area of declarative logic programming (rules). The application of rules was eventually revealed to be a valuable means of description thus further supporting the expressiveness of ontology representation [12][13]. Rules have been applied recently to extend OWL representations therefore being exposed to the same formal meanings of OWL-DL model-theoretic semantics [14]. More sophisticated implementations have been presented lately including the implementation of DLP-fusion which has been applied to the description of the bidirectional translation of premises and inference supporting two way translation between DL and DLP [15].

3 System Overview

Our development system begins with an initial ontology model. Next, our initial ontology model (with both TBOX and considerable ABOX data) is applied for input. The process begins with the identification of updated requirements for our model. These requirements support the development of rules for translation to a newer or more accurate representation. The rules are employed to support the following goals:

- Removal of existing properties, relationships and instances
- Inference of new rules to be updated
- Inference of new data translation
- Querying of data

In the use of updating class properties, the existing relationships must be removed. If the relationship maintains dependencies, the new updates need to be applied first including the identification of the new concepts. Once the modifications have taken place, then the instance data needs to be translated. In this step, existing identifiers are maintained so as to support the updated representations. In the instance modification, the data is then queried in order to test the output.

In cases where rules support updated restrictions, they are applied to support querying of the data sets before they are translated. The goal of these rules are to discern which portion of the ABOX representations meet the updated criterion as well as further discern categories of instance data that do not meet the criterion in order to expedite their repair. This process can result in a set of iterations in order to resolve the updates to enterprise-level ABOX instances resulting from the tightening of the constraint definition. After the data sets have been updated and verified to a more traditional step of verification is applied to the updated ontology. When these steps are completed, the ontology is updated and the rules are translated to an OWL-DL representation. This process is diagrammed in figure 1.

3.1 Implementation

Our implementation is based on the use of the Jena API, relying on its inference support as well as the provided built-in functions [16][17]. For support of a dynamic programming environment, we rely on the Jython language (Java based version of Python) as a means to develop within the Java-based Jena API [18][19]. This

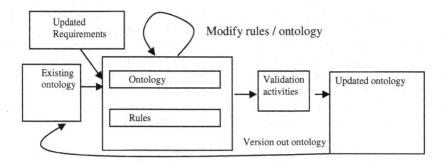

Fig. 1. System Overview Diagram

approach allows for dynamic modification of our deduced inference models within our model prototyping effort. Our system is employed through the application of Jython scripts which instantiate the Jena API into Jython-based objects (code example is presented below). The Jython-based approach allows for quick manipulation of the objects to support intermediate data generation beyond the scope of the Jena inference rule support and associated built-in functions. After an Ontology version is generated, rules are employed to query the Ontology. When the data is sufficiently verified, traditional Ontology validation methods supported by the Jena framework are applied.

```
from com.hp.hpl.jena.util import *
from com.hp.hpl.jena.rdf.model import *
from com.hp.hpl.jena.vocabulary import *
from com.hp.hpl.jena.reasoner import *
from com.hp.hpl.jena.reasoner.rulesys import *
PrintUtil.registerPrefix("eg","urn:x-hp:eg/")
m = ModelFactory.createDefaultModel()
configuration = m.createResource()
configuration.addProperty(ReasonerVocabulary.PROPruleMode,"forward")
configuration.addProperty(ReasonerVocabulary.PROPruleSet,"b.rules")
reasoner= GenericRuleReasonerFactory.theInstance().create(configuration)
data = FileManager.get().loadModel("file:CDex.xml")
infmodel = ModelFactory.createInfModel(reasoner,data)
id = infmodel.getDeductionsModel()
```

4   Case Study – Application to Development of an Engineering Ontology

This section presents abbreviated examples of rule sets that are applied to a subset of an ontology representing a bill of materials application. This RDF/OWL-based ontology exists as a layered application to an existing enterprise database. Our modifications reflect updates to the ontology design incorporating the Jena toolkit. These individual cases demonstrate the translation to the updated ontology to support updates to the existing design.

4.1   Inference of New Class and Property Definition

Our first example supports a change in requirements for a modification to incorporate two new (composite) object property relationships. The initial and updated portions of our ontology are diagrammed in figure 2 and figure 3. Each instance contains relevant data for the identification of each part type (Part number, part name, part type, cost, colors and effective date). They are grouped into a part group, associated by the means of an object property. The update consists of identifying a relationship from part groups to families to parts. The modifications to the original ontology is to the first remove the initial property relationship and establish two new object property relations defined as the part family and part group. Following the class definition, the rules support the inference of the defined property relationship.

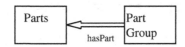

Fig. 2. Initial Ontology Relationship

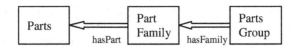

Fig. 3. Updated Ontology Relationship

The first rule (r1) removes earlier object properties relating the groups to the parts class, establishes the new relationships associating the classes then defines the part families instances.

```
[r1: (?b ?a  urn:x-hp:eg/Parts)  -> remove(0)
(urn:x-hp:eg/PartFamily rdf:type owl:Class)
(urn:x-hp:eg/hasFamily rdf:type owl:ObjectProperty)
(urn:x-hp:eg/hasFamily owl:domain urn:x-hp:eg/PartGroup)
(urn:x-hp:eg/hasFamily owl:range  urn:x-hp:eg/Parts)
(urn:x-hp:eg/hasPart owl:domain urn:x-hp:eg/PartFamily)
(urn:x-hp:eg/hasPart  owl:range  urn:x-hp:eg/PartGroup)
(urn:x-hp-eg/PartFamily_1   rdf:type   urn:x-hp:eg/Part_Family)
(urn:x-hp-eg/PartFamily_2   rdf:type   urn:x-hp:eg/Part_Family)
(urn:x-hp-eg/PartFamily_3 rdf:type urn:x-hp:eg/Part_Family)]
```

The rules r2-r7 associates the parts with the associated family groups also removing the earlier properties.

```
[r2: (?a urn:x-hp:eg/hasPart ?b) (?b urn:x-hp:eg/hasName urn:x-
hp:eg/screw)  ->  remove(0)  (urn:x-hp:eg/PartFamily_1  urn:x-hp-
eg:/hasPart ?b)]
[r3: (?a urn:x-hp:eg/hasPart ?b) (?b urn:x-hp:eg/hasName urn:x-
hp:eg/nuts)  ->  remove(0)  (urn:x-hp:eg/PartFamily_1  urn:x-hp-
eg:/hasPart ?b)]
[r4: (?a urn:x-hp:eg/hasPart ?b) (?b urn:x-hp:eg/hasName urn:x-
hp:eg/windshield) -> remove(0)  (urn:x-hp:eg/PartFamily_2 urn:x-
hp-eg:/hasPart ?b)]
[r5: (?a urn:x-hp:eg/hasPart ?b) (?b urn:x-hp:eg/hasName urn:x-
hp:eg/bolts)  ->  remove(0)  (urn:x-hp:eg/PartFamily_2  urn:x-hp-
eg:/hasPart ?b)]
```

```
[r6: (?a urn:x-hp:eg/hasPart ?b) (?b urn:x-hp:eg/hasName urn:x-
hp:eg/wheels) -> remove(0) (urn:x-hp:eg/PartFamily_3 urn:x-hp-
eg:/hasPart ?b)]
[r7: (?a urn:x-hp:eg/hasPart ?b) (?b urn:x-hp:eg/hasName urn:x-
hp:eg/seats) -> remove(0) (urn:x-hp:eg/PartFamily_3 urn:x-hp-
eg:/hasPart ?b)]
```

Rule 8 define the relations for group instances which maintain the associations with the respective families.

```
[r8: (?a rdf:type urn:x-hp:eg/partGroup) -> remove(0)
(urn:x-hp:eg/PartGroup_1        urn:x-hp-eg:/hasFamily        urn:x-
hp:eg/PartFamily_1)
(urn:x-hp:eg/PartGroup_2        urn:x-hp-eg:/hasFamily        urn:x-
hp:eg/PartFamily_2)
(urn:x-hp:eg/PartGroup_2        urn:x-hp-eg:/hasFamily        urn:x-
hp:eg/PartFamily_3)]
```

4.2 Restriction of Cardinality of a Specific Ontology Based Concept

In the second example, we examine a different portion of our ontology. This consists of the relationship between the object property constraints that identify the parts and usages (defined applications for a single part) classes. This example demonstrates the exploration of tightened constraints on the object property relationships between our two defined classes. The goal is to restrict the cardinality between these two objects so that they can maintain a 1:1 relationship between the instances. In the example rule (defined below) we rely on the built-in function notEqual to assist in the querying of the ontology in order to identify any instances that will be modified by the jena API. The use of notEqual() supports the identification of instances not supported in the existing ontology. When identified by the rule, the relationship is established in a separate class for verification. (figure 4,5)

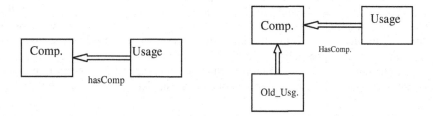

Fig. 4. Initial representation   **Fig. 5.** Updated Representation

```
[r1:    (urn:x-hp:eg/#Usage_1      urn:x-hp:eg/hasComponent      ?b)
notEqual(?b urn:x-hp:eg/Component_1) ->
remove(0)
(urn:x-hp:eg/EarlierUsages rdf:type owl:Class)
(urn:x-hp:eg/hasEarlierUsages rdf:type owl:ObjectProperty)
(urn:x-hp:eg/hasEarlierUsages          rdf:domain          urn:x-
hp:eg/EarlierUsages)
(urn:x-hp:eg/hasEarlierUsages rdf:range urn:x-hp:eg/Parts)
(urn:x-hp:eg/Component_1_OLD urn:x-hp:eg/hasEarlierUsages ?b)]
```

4.3 Inferring New Instance Classes Based on Specific Requirements

In our third example, we infer the existence of new instance classes. In this case, we derive a new class defined as a commodity part which exists as a property extended to the Parts definition. In this example, we rely on the Jena built-in functions to identify only those parts that cost less than 5.10 and higher in volume than 100K. Here, we are relying on the Jena greaterThan() and lessThan() function to establish new object properties that recognize specific constraints that define a commodity item (by being in a specific part family and under a specific cost level)(figure 6).

Fig. 6. Updated Commonality Relationship

```
[r1: (?a urn:x-hp:eg/partCost ?b)   (?a urn:x-hp:eg/partVolume ?d
)lessThan(?b 5.10) greaterThan(?d 100000.) ->
(urn:x-hp:eg/CommodityPart rdf:type owl:Class)
(urn:x-hp:eg/hasPart    rdf:domain    urn:x-hp:eg/CommodityParts)
(urn:x-hp:eg/hasPart     rdf:range      urn:x-hp:eg/Parts)(urn:x-
hp:eg/Commodity1 urn:x-hp:eg/hasParts ?b)]
```

5 Conclusion

A methodology and examples of rule technologies for the development of ontologies has been presented. This approach is primarily used to assist in support of integrating data to the Semantic Web. By the application of rules to infer new triple relationships, we have demonstrated a fast-prototyping approach to the advanced development of ontologies. Our environment is complementary to the traditional ontology extension and validation activities, focusing on implementing ontology updates on a more iterative level. Our employment of rules focuses on translation of primarily ABOX data, serving in a dual capacity of documenting logic between ontology versions. By employing the Jena API within the interactive Jython environment, we were able to support a dynamic approach to ontology development. Future opportunities in this area include the automatic generation of rules for purpose of translation of ontologies for greater support of bringing legacy applications and data to the Semantic Web.

Acknowledgements

Alan Fisk, Technical Leader of the Office of Technical Fellow, Ford Motor Company for insight provided in the development of this work.

References

1. Lehmann, T.: A Framework for Ontology based Integration of Structured IT-Systems, Workshop on Ontology-based Software Engineering. In: European Semantic Web Conference, Innsbruck, AU (2007)

2. Obitko, M.: Translation between Ontologies, Gerstner Laboratorory, Department of Cybernetics, Czech Technical University, http://www.webing.felk.cvut.cz/output/pub/WEBING-070504.pdf
3. Dou, D., McDermott, D., Qi, P.: Ontology Translation on the Semantic Web, http://cs-www.cs.yale.edu/homes/dvm/daml/ontomerge_odbase.pdf
4. Noy, N.F., McGiunness, D.L.: Ontology Development 101: Guide to Creating Your First Ontology, http://protege.stanford.edu/publications/ontology_development/ontology101-noy-mcguinness.html
5. Powers, S.: Practical RDF, Oreilly pub. (July 2003)
6. Knublauch, H.: Semantic Web Rules – Tools and Languages, tutorial. In: RuleML 2006 (2006), http://2006.ruleml.org/slides/tutorial-holger.pdf
7. Reynolds, D.: JUC-Jena Rules, HP Laboratories, http://jena.hpl.hp.com/juc2006/proceedings/reynolds/rules-slides.ppt
8. Guo, Y.-k., Darlington, J.: The Unification of Functional and Logic Languages-Towards Constraint Functional Programming, IEEE (1989)
9. Bechhofer, S., van Harmelen, F., Hendler, J., McGuinness, D., Patel S., Peter F., Stein, L.A.: OWL Web Ontology Language Reference, W3C (February 2004), http://www.w3.org/TR/2004/REC-owl-ref-20040210
10. Smith, M.K., Welty, C., McGiunness, D.: W3C OWL Web Ontology Language, http://www.w3.org/TR/owl-guide/
11. Alexander, B.: Description Logics in Data Management. IEEE Transactions on Knowledge and Data Engineering 7(5) (October 1995)
12. Alsac, G., Baral, C.: Reasoning in description logics in declarative logic programming, AAAI (2002)
13. Motik, B., Rosati, R., Sattler, U.: Can OWL and Logic Programming Live Together Happily Ever After? In: ISWC 2006, http://iswc2006.semanticweb.org/items/Motik2006bh.pdf
14. Horrocks, I., Patel-Schneider, P.F., Bechhofer, S., Tsarkov, D.: OWL Rules: A Proposal and Prototype Implementation
15. Grosof, B.N., Raphael, V., Horrocks, I., Decker, S.: Description Logic Programs: Combining Logic Programs with Description Logic. In: WWW 2003, Budapest, Hungary (May20-24, 2003)
16. Jena: A Semantic Web Framework For Java, http://jena.sourceforge.net/
17. http://jena.sourceforge.net/inference/
18. http://www.jython.org
19. Pedroni, S., Rappin, N.: Jython Essentials. O'Reilly pub. (2002)

Integrating Rules and Description Logics with Circumscription for the Semantic Web

Fangkai Yang[1] and Xiaoping Chen[2]

Multi-Agent System Lab
University of Science and Technology of China, Hefei, China
fkyang@mail.ustc.edu.cn, xpchen@ustc.edu.cn

Abstract. In this paper, we propose $\mathcal{DL}clog$, a new hybrid formalism combining Description Logics(DL) and Logic Programming(LP) for Semantic Web serving as an extension to $\mathcal{DL} + log$[17]. By $\mathcal{DL}clog$, users can reason nonmonotonically with DL ontology. We introduce negative dl-atoms to the bodies of the rules, and extend Nonmonotonic Semantics (NM-Semantics) of $\mathcal{DL} + log$ to evaluate dl-atoms with *circumscriptive models* of DL ontology in the sense of parallel circumscription. In this way, negative dl-atoms are treated nonmonotonically, while the formalism still remains faithful to NM-Semantics, DL and LP. We also present a decision procedure for the extended semantics based on a restricted form of $\mathcal{DL}clog$, in which DL ontologies are written with $\mathcal{ALCIO}$ or $\mathcal{ALCQO}$ and roles are not allowed to occur in negative dl-atoms.

1 Introduction

The problem of adding rules to Description Logics is currently a hot research topic, due to the Semantic Web applications of integrating rule-based systems with ontologies. Practically, most research work[14,15,16,17] in this area focuses on the integration of Description Logic with datalog rules or its non-monotonic extensions, and $\mathcal{DL}+log$[17] is a powerful result of a series of hybrid approaches: DL-log[14], r-hybrid KB[16], r^+-hybrid KB[15], which define integrated models on the basis of single models of classical theory[4].

However, there is a severe limitation in $\mathcal{DL} + log$ that DL predicates cannot occur behind "not" in rules. This syntactical restriction makes it impossible to use rules to draw conclusions by the results currently underivable from DL ontology, which cannot satisfies practical needs such as closed world reasoning and modeling exceptions and defaults for DL predicates. To overcome this limitation requires to introduce negative dl-atoms in the body of the rules and interpret them with a nonclassical semantics, which is a nontrivial generalization of Nonmonotonic Semantics (NM-Semantics) of $\mathcal{DL} + log$. As DL adopts Open World Assumption(OWA), we must introduce a kind of closed world reasoning to DL ontology to interpret the unknown results as negation. Such closed world reasoning must be transparently integrated into the framework of NM-Semantics in order that the generalized semantics remain faithful to NM-Semantics, DL and LP. Finally, we hope the definition of the semantics is concise, model-theoretical and if possible, decidable.

A. Paschke and Y. Biletskiy (Eds.): RuleML 2007, LNCS 4824, pp. 182–189, 2007.

In this paper, a new hybrid formalism $\mathcal{DL}clog$ is presented to achieve these goals. In syntax, we allow DL predicates occurring behind "not" in rules, and in semantics, extend NM-Semantics of $\mathcal{DL} + log$ to be Nonmonotonic Circumscriptive Semantics(NMC-Semantics). In NMC-Semantics, dl-atoms in rules are evaluated under the *circumscriptive models* of the DL ontology in the sense of McCarthy's parallel circumscription[9] rather than classical models as in NM-Semantics. Parallel circumscription is a general form of circumscription formalizing Closed World Assumption(CWA)[13], and is equivalent to Extended CWA[7]. By parallel circumscription, "not" in negative dl-atoms are interpreted in a nonmonotonic manner. As circumscriptive models serve as intermediate models only used to evaluate dl-atoms, NMC-semantics remains faithful to DL, LP and NM-Semantics of $\mathcal{DL} + log$, in that users can switch $\mathcal{DL}clog$ KB to any of these formalisms by syntactical restriction.

We leave the decidability of $\mathcal{DL}clog$ with respect to NMC-satisfiability as an open problem, due to the open problem of decidability of *concept-fixing cKBs* in [2]. A restricted form of $\mathcal{DL}clog$ is decidable with the complexity of NExP^{NP}-complete, in which ontologies are written in $\mathcal{ALCIO}$ or $\mathcal{ALCQO}$ and roles are not allowed to occur under "not" in rules. For this form, we present a decision procedure.

The rest of the paper is organized as follows. Section 2 presents the syntax and semantics of $\mathcal{DL}clog$. Section 3 presents the algorithms for reasoning with $\mathcal{DL}clog$ including decidability and complexity results. Related work is presented in Section 4 and Section 5 ends with conclusion and future work. For the limitation of space, we refer [9] for the reader unfamiliar with circumscription, and [17] for $\mathcal{DL} + log$. We recommend a full version of our paper[18] to interested readers.

2 $\mathcal{DL}clog$: Syntax and Semantics

- **Syntax**

In the syntax of $\mathcal{DL}clog$, we generalize $\mathcal{DL} + log$ with negative dl-atoms, and restricts the rules to be DL-safeness[10] rather than weak-safeness. The alphabet of predicates Σ is divided into three mutually disjoint sets $\Sigma = \Sigma_C \cup \Sigma_R \cup \Sigma_D$, where Σ_C is an alphabet of concepts names, Σ_R is an alphabet of role names and Σ_D is an alphabet of Datalog predicates.

Definition 1. *A hybrid knowledge base $\mathcal{K}$ is a pair $(\mathcal{O}, \mathcal{P})$, where $\mathcal{O}$ is the a description logic ontology $(\mathcal{T}, \mathcal{A})$ as its TBOX and ABOX[1], and $\mathcal{P}$ is a finite set of Datalog$^{\neg, \vee}$ rules of following form:*

$$\mathcal{R}: \ H_1(X_1) \vee \ldots \vee H_k(X_h) \leftarrow RB_1(Y_1), \ldots, RB_m(Y_m),$$
$$not \ RB_{m+1}(Y_{m+1}), \ldots, not \ RB_s(Y_s), CB_1(Z_1), \ldots, CB_n(Z_n),$$
$$not \ CB_{n+1}(Z_{n+1}), \ldots, not \ CB_t(Z_t).$$

where $H_i(X_i)$, $RB_i(Y_i)$, $CB_i(Z_i)$ are atoms, X_i, Y_i, and Z_i are vectors of variables. Let $\mathcal{C}$ denote a set of countably infinite constant names. And

- *each H_i is either a DL predicate or a Datalog predicate.*
- *each RB_i is a Datalog predicate, and $RB_i(Y_i)$ is called a rule-atom.*

- each CB_j is a DL predicate, $CB_j(Z_j)$ is called a dl-atom.
- (DL-safeness)every variable of $\mathcal{R}$ must appear in at least one of the $RB_i(Y_i)$($1 \leq i \leq m$).

We use $M(\mathcal{P})$ to denote the set of DL predicates occurring under "not" in $\mathcal{P}$. $atom_C^+(\mathcal{R}) = \{CB_i(Z_i)|1 \leq i \leq n\}$ denotes all dl-atoms occurring positively in $\mathcal{R}$. $atom_C^-(\mathcal{R}) = \{CB_j(Z_j)|n+1 \leq j \leq t\}$ denotes all the dl-atoms occurring under "not" in $\mathcal{R}$. $atom_C(\mathcal{R}) = atom_C^+(\mathcal{R}) \cup atom_C^-(\mathcal{R})$ denotes the DL-atoms in $\mathcal{R}$. Besides, we use $CIRC[A; P]$ to denote the parallel circumscription $CIRC[A; P; Z]$ with $Z = \emptyset$, indicating that all other predicates' interpretations can vary to support the minimization.

- **Semantics**

Like $\mathcal{DL} + log$, in the definition of the semantics of $\mathcal{DL}clog$ we adopt Standard Names Assumption(SNA): every interpretation is defined over the same fixed, countably infinite domain Δ, and the alphabet of constant $\mathcal{C}$ is such that it is in the same one-to-one correspondence with Δ in each interpretation. For a program $\mathcal{P}$ and a set of constants $\mathcal{C}$, we use $gr(\mathcal{P}, \mathcal{C})$ to denote the program obtained from $\mathcal{P}$ by substituting all the variables in $\mathcal{P}$ with all possible constants in $\mathcal{C}$. In order to treat negative dl-atoms, we use following projection to reduce dl-atoms in rules relative to an interpretation.

Definition 2. *Given an interpretation $\mathcal{I}$ over an alphabet of predicates $\Sigma' \subset \Sigma$ and a ground program $gr(\mathcal{P}, \mathcal{C})$ over the predicates in Σ, the projection of $gr(\mathcal{P}, \mathcal{C})$ with respect to $\mathcal{I}$, denoted by $\Pi(gr(\mathcal{P}, \mathcal{C}), \mathcal{I})$ is the program obtained from $gr(\mathcal{P}, \mathcal{C})$ as follows. Let $r(t)$ denote a literal (either unary or binary) in $gr(\mathcal{P}, \mathcal{C})$, for each rule $\mathcal{R} \in gr(\mathcal{P}, \mathcal{C})$,*

- *delete $\mathcal{R}$ if there exists an atom $r(t)$ in the head of $\mathcal{R}$ such that $r \in \Sigma'$ and $t \in r^{\mathcal{I}}$;*
- *delete each atom $r(t)$ in the head of $\mathcal{R}$ such that $r \in \Sigma'$ and $t \notin r^{\mathcal{I}}$;*
- *delete $\mathcal{R}$ if there exists a positive literal $r(t)$ in the body of $\mathcal{R}$ such that $r \in \Sigma'$ and $t \notin r^{\mathcal{I}}$;*
- *delete each positive literal $r(t)$ in the body of $\mathcal{R}$ such that $r \in \Sigma'$ and $t \in r^{\mathcal{I}}$;*
- *delete $\mathcal{R}$ if there exists a negative atom not $r(t)$ in the body of $\mathcal{R}$ such that $r \in \Sigma'$ and $t \in r^{\mathcal{I}}$;*
- *delete each negative atom not $r(t)$ in the body of $\mathcal{R}$ such that $r \in \Sigma'$ and $t \notin r^{\mathcal{I}}$;*

Based on this definition, dl-atoms can be eliminated by being projected with an interpretation on $\Sigma_R \cup \Sigma_C$. Then one can obtain its stable model via *Gelfond-Lifschitz Transformation*[8]. In following we define the Nonmonotonic Circumscriptive Semantics(NMC-Semantics) for $\mathcal{DL}clog$.

Definition 3. *(Nonmonotonic Circumscriptive Semantics, NMC-Semantics)For a hybrid KB $\mathcal{K} = (\mathcal{O}, \mathcal{P})$ and $\mathcal{C}$ the set of individuals explicitly stated in $\mathcal{O}$, let $\mathcal{U}, \mathcal{V}, \mathcal{W}$ be sets of interpretations on language $\Sigma_C \cup \Sigma_R$, $\Sigma_C \cup \Sigma_R$ and Σ_D, respectively. A structure $\mathcal{M} = (\mathcal{U}, \mathcal{V}, \mathcal{W})$ is the Nonmonotonic Circumscriptive Model (NMC-Model) of $\mathcal{K}$, denoted as $\mathcal{M} \models_{NMC} \mathcal{K}$, if and only if*

- *for each $\mathcal{I} \in \mathcal{U}$, $\mathcal{I} \models \mathcal{O}$.*
- *for each $\mathcal{I} \in \mathcal{V}$, $\mathcal{I} \models CIRC[\mathcal{O}; M(\mathcal{P})]$.*
- *for each $\mathcal{J} \in \mathcal{W}$, $\mathcal{J}$ is a stable model of $\Pi(gr(\mathcal{P}, \mathcal{C}), \mathcal{I}_c)$ where $\mathcal{I}_c \in \mathcal{V}$.*

We call $\mathcal{U}, \mathcal{V}, \mathcal{W}$ the classical part, circumscriptive part, and stable part of the NMC-model. $\mathcal{K}$ is NMC-satisfiable if and only if it has an NMC-model without any part as $\emptyset$. c denotes a tuple of constants. A ground atom $p(c)$ is NMC-entailed by $\mathcal{K}$, denoted as $\mathcal{K} \models_{NMC} p(c)$, if and only if

- *if p is a DL predicate, for each interpretation $\mathcal{I} \in \mathcal{U}$, $\mathcal{I} \models p(c)$.*
- *if p is a rule predicate, for each interpretation $\mathcal{J} \in \mathcal{W}$, $\mathcal{J} \models p(c)$.*

We present an examples to illustrate such semantics.

Example 1. Given following $\mathcal{DL}clog$ KB $\mathcal{K} = (\mathcal{O}, \mathcal{P})$.

$\mathcal{O}$	$\mathcal{P}$
$\neg seaside \equiv notSC$	$seasideCity(x) \leftarrow portCity(x), O(x), not\ notSC(x)$
$portCity(Barcelona)$	$O(Barcelona)$

We analyze following three cases.

1. Querying $seasideCity(Barcelona)$. With NMC-Semantics, negative dl-atom "$not\ notSC(x)$" is satisfied in circumscriptive models containing $\neg notSC$ $(Barcelona)$, in which $notSC$ is circumscribed. This is the only model used for evaluating dl-atoms. Thus we obtain $\mathcal{K} \models_{NMC} seasideCity(Barcelona)$.
2. Query $notSC(Barcelona)$. $\mathcal{O}$ have two models $\mathcal{I}_1$ and $\mathcal{I}_2$, where $notSC(Barcelona) \in \mathcal{I}_1$ and $notSC(Barcelona) \notin \mathcal{I}_2$. $\mathcal{I}_2$ is a circumscriptive model. Thus we have $\mathcal{O} \nvDash_{NMC} notSC(Barcelona)$. We can see that in NMC-Semantics, circumscriptive models don't affect reasoning with $\mathcal{O}$, which remains to be monotonic and classical.
3. Circumscriptive models can introduce nonmonotonicity to reasoning with rules by negative dl-atoms, Once we add "$notSC(Barcelona)$" into $\mathcal{O}$, we obtain $\mathcal{K} \models_{NMC} \neg seasideCity(Barcelona)$.

• **Semantic Properties**

NMC-Semantics is the generalization of $\mathcal{DL} + log$. When there are no negative dl-atoms, $\mathcal{DL}clog$ is reduced to $\mathcal{DL} + log$ in both syntax and semantics, as is stated in following proposition.

Proposition 1. *For a $\mathcal{DL}clog$ KB $\mathcal{K} = (\mathcal{O}, \mathcal{P})$, when there are no negative dl-atoms in $\mathcal{P}$, $\mathcal{K}$ is NMC-satisfiable iff $\mathcal{K}$ is NM-satisfiable in the sense of $\mathcal{DL}+log$.*

NMC-Semantics is also faithful to DL and LP, in that DL and LP are the restricted forms of $\mathcal{DL}clog$.

Proposition 2. *For a $\mathcal{DL}clog$ KB $\mathcal{K} = (\mathcal{O}, \mathcal{P})$, (i) if $\mathcal{P} = \emptyset$, $\mathcal{K}$ is NMC-satisfiable iff $\mathcal{O}$ is classically satisfiable. (ii) if $\mathcal{O} = \emptyset$, $\mathcal{K}$ is NMC-satisfiable iff $\mathcal{P}$ has stable model(s).*

Besides, we need following comments about NMC-Semantics.

1. **Faithfulness.** The above two proposition shows that $\mathcal{DL}clog$ is a sound extension to $\mathcal{DL}+log$, and remains to be faithful to NM-Semantics of $\mathcal{DL}+log$, DL and LP, in that the users can switch a $\mathcal{DL}clog$ to each of these formalisms by imposing syntax restrictions. When a KB is syntactically switched to any of the formalisms, the semantics also varies accordingly.
2. **Nonmonotonicity.** NMC-Semantics is nonmonotonic. Besides the nonmonotonic nature of negative rule-atoms, negative dl-atoms, which are evaluated by circumscriptive models, also add nonmonotonicity features to rules. This comment is verified in the third case of the above example.
3. **OWA vs. CWA.** As circumscriptive model is only an intermediate part used to evaluate dl-atoms, it does not affect any semantic feature of DL. As a result, semantics of DL adopts OWA. By contrast, the models for the rule atoms are obtained as stable model, in which OWA is adopted.
4. **UNA vs. Non-UNA.** To abridge the gap between Unique Names Assumption (UNA) of DL and Non-Unique Names Assumption of LP, we inherit the Standard Names Assumption (SNA) and impose it to DL. As SNA implies UNA, such strategy solves this contradiction.

Consequently, NMC-Semantics fulfils our goal in Section 1 with a concise, model-theoretical semantics, and provide desirable semantic properties.

3   Reasoning with $\mathcal{DL}clog$

In this section, we present an algorithm for a restricted form of $\mathcal{DL}clog$: roles are not allowed to occur under "not" in rules, and $\mathcal{O}$ is written in $\mathcal{ALCIO}$ or $\mathcal{ALCQO}$. We can obtain a sound and complete decision procedure for this restricted form. For general form of $\mathcal{DL}clog$, computing the circumscription involves *concept-fixing cKB* and more expressive DLs, whose decidability is left as an open question till [2]. Our algorithm is based on two existing algorithms: the r-hybrid KB's reasoning algorithm in [16] and the algorithm to compute circumscription of concept-circumscribed cKBs[2]. Firstly, we introduce the formal notations.

Given a $\mathcal{DL}clog$ $\mathcal{K} = (\mathcal{O}, \mathcal{P})$, we use $gr_p(\mathcal{P})$ to denote all the ground atoms in $gr(\mathcal{P}, \mathcal{C})$. For the set of ground atoms $gr_p(\mathcal{P})$, we use (G_P, G_N) to denote its partition. For a set of ground atoms S, use $\neg S$ to denote $\{\neg m(t)|m(t) \in S\}$. Define $\mathcal{P}(G_P, G_N)$ as

$$\mathcal{P}(G_P, G_N) = \mathcal{P} \cup G_P \cup \{\leftarrow r(x)|r(x) \in atom_C(\mathcal{P})\}$$

Finally, define $G_{PC} = gr(atom_C^-(\mathcal{P}), \mathcal{C}) \cap G_P$, $G_{NC} = gr(atom_C^-(\mathcal{P}), \mathcal{C}) \cap G_N$, and furthermore, $\bigwedge G_{PC} = \bigwedge_{a \in G_{PC}} a$, and respectively, $\bigwedge \neg G_{NC} = \bigwedge_{a \in G_{NC}} \neg a$.

The following is the notion of ground atoms induced by an interpretation.

Definition 4. *(Definition 6,[16])Let $\mathcal{I}$ be an interpretation and (G_P, G_N) be the partition of $gr_p(\mathcal{P})$ such that for each $r(t) \in gr_p(\mathcal{P})$, $t \in r^{\mathcal{I}}$ if and only if $r(t) \in G_P$. We call (G_P, G_N) the partition of $gr_p(\mathcal{P})$ induced by $\mathcal{I}$.*

Furthermore,we introduce counting formulae as a common generalization of TBOX and ABOX.

Definition 5. *(Definition 7,[2]) A counting formula ϕ is a boolean combination of GCIs(General Concept Inclusions), ABox assertions $C(a)$ and cardinality assertions $(C = n)$ and $(C \leq n)$ where C is a concept and n is a non-negative integer. We use $\vee$, $\wedge$, $\neg$ and $\rightarrow$ to denote the boolean operators of counting formulae. An interpretation $\mathcal{I}$ satisfies a cardinality assertion $(C = n)$ if $|C^{\mathcal{I}}| = n$(the cardinality of the domain of C), and $(C < n)$ if $|C^{\mathcal{I}}| < n$. The satisfaction relation $\mathcal{I} \models \phi$ between models $\mathcal{I}$ and counting formulae ϕ is defined in obvious way.*

We use $C \sqsubset D$ as an abbreviation of $(C \sqsubseteq D) \wedge \neg(D \sqsubseteq C)$. For a set of predicates $M(\mathcal{P}) = \{A_0, \dots, A_n\}$, given a subset $S \subseteq M(\mathcal{P})$, define

$$C_S = \bigsqcap_{A \in M(\mathcal{P})} A \sqcap \bigsqcap_{A \in \{A_1, \dots, A_n\} - S} \neg A$$
$$\xi_1 = \bigwedge_{A \in M(\mathcal{P})} (\neg(A' \sqsubseteq A) \rightarrow \bigvee_{B \in M(\mathcal{P})} (B \sqsubset B'))$$
$$\xi_2 = \bigvee_{A \in M(\mathcal{P})} (A' \sqsubset A) \wedge \bigwedge_{B \in M(\mathcal{P})} (B \equiv B')$$

For each concept C and role R, introduce fresh names C' and R'. For a concept B, denote B' the result by replacing each occurrence of C with C' and R with R'. $\mathcal{O}'$ can be defined analogously. By [2], the concept is satisfiable in $CIRC[\mathcal{O}; M(\mathcal{P})]$ if and only if it is satisfiable in a finite model, and the algorithm determining the satisfiability of counting formulae in $\mathcal{ALCIO}$ and $\mathcal{ALCQO}$ is the upper bound NExp. We obtain following algorithm.

> **Algorithm** NMCSat-Restricted($\mathcal{O}, \mathcal{P}$)
> **Input:** DL KB $\mathcal{O}$ and a set of rules $\mathcal{P}$
> **Output:** true if $(\mathcal{O}, \mathcal{P})$ is NMC-satisfiable, and false otherwise
> **begin**
> if there exist
>   - a partition (G_P, G_N) of $gr_p(\mathcal{P})$
>   - a sequence $\{n_S | S \subseteq M(\mathcal{P})\}$ of non-negative integers in binary
> such that following conditions are satisfied
> 1. $\mathcal{P}(G_P, G_N)$ has stable model
> 2. first order theory $\mathcal{O} \cup G_P \cup \neg G_N$ is satisfiable
> 3. Oracle tells that ϕ_1 is satisfiable, where
> $$\phi_1 = \bigwedge_{p \in \mathcal{O}} p \wedge \bigwedge_{S \subseteq M(\mathcal{P})} (C_S = n_S) \wedge \bigwedge G_{PC} \wedge \bigwedge \neg G_{NC}$$
> 4. Oracle tells that ϕ_2 is unsatisfiable, where
> $$\phi_2 = \bigwedge_{p \in \mathcal{O}'} p \wedge \bigwedge_{S \subseteq M(\mathcal{P})} (C_S = n_S) \wedge \bigwedge G_{PC} \wedge \bigwedge \neg G_{NC} \wedge \xi_1 \wedge \xi_2.$$
> returns true else return false.
> **end**

We have following proposition stating the termination of the algorithm.

Proposition 3. *NMCSat-Restricted($\mathcal{O}, \mathcal{P}$) terminates within finite time.*

Following result states the correctness of the algorithm.

Proposition 4. *For a $\mathcal{DLclog}$ KB $\mathcal{K} = (\mathcal{O}, \mathcal{P})$, if $\mathcal{O}$ is written with $\mathcal{ALCIO}$ or $\mathcal{ALCQO}$, and roles are not allowed to occur under "not" in $\mathcal{P}$, it is NMC-satisfiable if and only if NMCSat-Restricted($\mathcal{O}, \mathcal{P}$) returns true.*

For the complexity issues, we remind of the exponential hierarchy NExP$\subseteq$ NP^{NEXP} $\subseteq$ NExP^{NP} $\subseteq$2-ExP. Thus we have the complexity of deciding NMC-satisfiability of the restricted $\mathcal{DL}clog$.

Proposition 5. *For a $\mathcal{DL}clog$ KB $\mathcal{K} = (\mathcal{O}, \mathcal{P})$, if $\mathcal{O}$ is written with $\mathcal{ALCIO}$ and $\mathcal{ALCQO}$, and roles are not allowed to occur under "not" in $\mathcal{P}$, NMC-satisfiability can be decided in the complexity* NEXP^{NP}*-complete.*

4   Related Work

The first formal proposal for integrating Description Logics and rules is $\mathcal{AL}$-log[5]. This proposal is generalized in [14] with Datalog$^{\neg,\vee}$ and usage of roles. This series of work was continued in [16], named r-hybrid KB, where DL-safety[10] condition and Standard Names Assumption is firstly adopted, and nomonotonic negation is introduced into the body of the rules. It has two subsequent research work. In [15], the author pointed out Standard Names Assumption can be substituted with Non-UNA for both DL and rules. In [17], r-hybrid is generalized to be $\mathcal{DL} + log$ in an other direction, where DL-safety is substituted by weak safeness, and thus existential entailment is possible with $\mathcal{DL} + log$. We inherit the framework of these methods. By introducing negative dl-atoms and using parallel circumscription to evaluate them, we obtain the semantics with nonmonotonic features treating negative dl-atoms. Compared with previous work, this feature is novel and important, in that negative DL-atoms can be truly nonmonotonically interpreted.

Recently, some "full-integration" methods are proposed. [3] proposed to use First Order Autoepistemic Logic[6] as a host language to accommodate DL and rules. [11] proposed to build a hybrid KB in the framework of the logic of MKNF. Both of the methods treat DL and LP in a uniform logic rather than integrating existing formalisms. Compared with these work, instead of extending language, our formalism is based on a hybrid, modular semantics integrating classical semantics, circumscription and stable model. $\mathcal{DL}clog$ makes it possible to use existing reasoning engines of DL and LP to build a practical system.

5   Conclusion and Future Work

In this paper, we present a hybrid formalism $\mathcal{DL}clog$ as both semantic and syntactic extension of Rosati's $\mathcal{DL} + log$ by allowing negative dl-atoms occurring in the body of the rules. To obtain the stable models of the rules, dl-atoms are evaluated by the circumscriptive models of the DL ontology in the sense of parallel circumscription. In this way, the negative dl-atoms are treated nonmontonically and is closely similar to the treatment of "not" in LP. This formalism strengthens the nonmonotonic expressing and reasoning ability of $\mathcal{DL} + log$, and remains faithful to the NM-Semantics of $\mathcal{DL} + log$, DL and LP. Besides, we present a decision procedure when roles do not occur under "not" in rules and ontologies are written in $\mathcal{ALCIO}$ and $\mathcal{ALCQO}$.

As our future work, we will focus on deterministic algorithm which is used to reason with $\mathcal{DL}clog$ in a prototype system.

References

1. Baader, F., Calvanese, D., McGuinness, D., Nardi, D., Patel-Schneider, P. (eds.): The Description Logic Handbook. Cambridge University Press, Cambridge (2003)
2. Bonatti, P., Lutz, C., Wolter, F.: Description Logics with Circumscription. In: KR 2006 (2006)
3. Bruijn, J., Eiter, T., Polleres, A., Tompits, H.: Embedding Non-Ground Logic Programs into Autoepistemic Logic for Knowledge-Base Combination. In: Proc of IJCAI 2007 (2007)
4. Bruijn, J., Eiter, T., Polleres, A., Tompits, H.: On Representational Issues About Combinations of Classical Theories with Nonmonotonic Rules. In: Lang, J., Lin, F., Wang, J. (eds.) KSEM 2006. LNCS (LNAI), vol. 4092, pp. 1–22. Springer, Heidelberg (2006)
5. Donini, F., Lenzerini, M., Nardi, D., Schaerf, W.: $\mathcal{AL}$-log: Integrating Datalog and Description Logics. Artificial Intelligence 100(1-2), 225–247 (1998)
6. Moore, R.: Semantical considerations on nonmonotonic logic. Artificial Intelligence 25(1), 75–94 (1985)
7. Gelfond, M., Przymusinska, H., Przymusinki, T.: On the Relationship between Circumscription and Negation as Failure. Artificial Intelligence 38, 75–94 (1989)
8. Gelfond, M., Lifschitz, V.: Classical Negation in Logic Programs and Disjunctive Databases. New Generaltion Computing 9, 365–385 (1991)
9. McCarthy, J.: Applications of circumscription in formalizing common sense knowledge. Artificial Intelligence 28, 89–116 (1986)
10. Motik, B., Sattler, U., Studer, R.: Query answering for OWL-DL with rules. In: McIlraith, S.A., Plexousakis, D., van Harmelen, F. (eds.) ISWC 2004. LNCS, vol. 3298, Springer, Heidelberg (2004)
11. Motik, B., Rosati, R.: A Faithful Integration of Description Logics with Logic Programming. In: Proc. of IJCAI 2007 (2007)
12. Lifschitz, V.: Nonmonotonic Databases and Epistemic Queries. In: Proc. IJCAI 1991 (1991)
13. Reiter, R.: On Closed-world Data Bases. In: Gallaire, H., Minker, J. (eds.) Logic and Data Bases, pp. 55–76. Plenum Press, New York (1978)
14. Rosati, R.: Towards expressive KR systems integrating Datalog and description logics: Preliminary report. In: Proc. of DL 1999 (1999)
15. Rosati, R.: Semantic and Computational Advantages of the Safe Integration of Ontologies and Rules. In: Fages, F., Soliman, S. (eds.) PPSWR 2005. LNCS, vol. 3703, pp. 50–64. Springer, Heidelberg (2005)
16. Rosati, R.: On the decidability and complexity of integrating ontologies and rules. J. Web Semantics 3(1), 61–73 (2005)
17. Rosati, R.: $\mathcal{DL}+log$: Tight integration of description logics and disjunctive datalog. In: Proc. of KR 2006 (2006)
18. Yang, F., Chen, X.: Integrating Rules and Description Logics with Circumscription for Semantic Web (Techinical Report), in http://mail.ustc.edu.cn/ fkyang/ dlclog-full.pdf

XML Data Compatibility from the Ground Up

Karthick Sankarachary

karthicks@email.com

Abstract. While XML may have emerged as the de facto format of data exchanged by peers, it faces the exact same evolutionary challenges that has plagued preceding lingua francas. As peers grow organically, they will inevitably generalize or specialize the XML messages they give and take, at the risk of breaking down existing relationships. If anything, change-induced outages are more likely to occur due to XML, since it promotes loose-coupling. To ensure that peers are interoperable, we lay down some ground rules of compatibility, by generally applying set theory on the extensions of the OWL classes/properties representing XML concepts/relations. Specifically, we take a hand-in-glove approach where the statements of compatibility (based on RIF-esque rules) tag along as corollaries to those of XML (based on OWL ontologies). Along the lines of two-pass compilers, we first parse the XML Infoset into an intermediate OWL representation, prior to analyzing the semantics therein.

1 Introduction

Since time immemorial, the exchange of ideas between humans has required their comprehension of a shared vocabulary. In this day and age of web agents and enterprise systems, machines endeavour to emulate this human behavior using the same "design principles". In other words, for two or more peer applications to exchange messages in mutually beneficial ways, a prerequisite is that they use the same or similar vocabularies. In general, machine-readable vocabularies help to specify constraints on message data formats and formalize operational aspects such as service level agreements and abstract process definitions. At any rate, the circle of peers with whom one can make conversation will grow to the extent that one's shared vocabularies are open. In this sense, web standards have an important role to play in opening up the lines of peer-to-peer communication. The common thread across all existing and emerging web standards is undoubtedly XML [1], which specifies ways to organize information hierarchically as per well-formed schemas. Although XML data is mostly passive, it may contain actionable elements that drive lifecycles of services on the receiving end.

2 The Chaotic System of Forces

In our opinion, the pervasive use of XML as an open standard may be a double-edged sword. In part, this is due to the fact that it promotes loosely coupled development of peers, which in turn permits them to evolve fairly independently,

A. Paschke and Y. Biletskiy (Eds.): RuleML 2007, LNCS 4824, pp. 190–198, 2007.

even to the extent that they no longer understand each other. Unless the ensuing chaos is controlled, it may open the flood gates to a virtual strike of sorts. This calls for certain checks and balances to be put in place, in order to avoid disconnect and discontent within the network of peers. To that end, we describe herein a system of rules that will enable peers to establish and maintain working relationships with each other. In brief, the problem can be stated as follows:

Problem 1. Given a pair of peers, how can one analyze the gaps, if any, between the message formats they understand? Further, what are the implications if the pair is related historically in time (i.e. they're different versions of the same entity) versus physically in space (i.e. they co-exist and interact with each other.)

Herein, we limit the scope of our analysis to the data-specific parts of messages, specifically those for which an XML Schema [2] exists. In reality, message payloads may be service-oriented in the sense that they conform to a higher-order schema such as that of WSDL [3]. Even so, service providers typically act on and give back XML documents, hence our rules may still be applicable to some degree. In fact, we will describe later how the proposed data-centric approach can potentially be reused for other meta models, of higher orders of magnitude.

Solution 1. To begin with, we describe some of the semantic web technologies that we will use to capture our rules. Then, we nail down the essence of compatibility, in such a way that it is generic and hence reusable. Subsequently, we go over an approach that tries to make sense of an XML Schema by removing the syntactic sugar yet retaining its core essence. Finally, we apply our theories to all of the basic and advanced constructs in XML. Simultaneously, we define what it means for each of those constructs to be compatible as we go along.

3   The Nomenclature of Rules

The driving force of the semantic web initiative is OWL [4]. In brief, it specifies a way of defining vocabularies, in terms of concepts of kindred souls (a.k.a. classes) and relations between kindred concepts (a.k.a. properties.) In other words, a class defines a set of individuals (i.e. its extent) deemed to have certain characteristics (i.e. its properties.) Specifically, an extent may be defined either in full or in part (based on what is known), by name (of constituent individuals) or by reuse (of other classes). In general, a property conveys a one-way relationship from a domain to a range class, but it may itself take on characteristics such as symmetry, transitivity, uniqueness and referential integrity.

While in general it is possible to express rules in terms of OWL axioms, at times it falls short, such as when the situation calls for full first-order logic (e.g. generic universal & existential quantification) and on-the-fly conceptualization (viz. when the classes and properties are not known a priori). Specifically, our requirements are such that the rule language must comply with first-order logic (FOL) yet interoperate with other languages (including OWL). For the sake of usability, it must be not only exceptionally expressive but also human readable.

While there a number of rule languages out there, the ones that come closest to OWL are RIF [5], SWRL [6] and N3 [7]. Whereas SWRL tends to redefine rather than extend OWL, and whereas N3 lacks somewhat in formalism, RIF appears to be headed in the right direction although it is itself very much a work in progress. Given RIF's charter to be interchangeable with like-minded languages, we opt to align ourselves with RIF. However, we felt it necessary to take the following liberties with RIF for the sake of readability/completeness:

1. OWL descriptions and properties may be denoted as $C(x)$ and $P(x\ y)$ respectively, a la SWRL (e.g. *owl:equivalentClass(owl:Class(?x) owl:Class(?x))*).
2. RIF existential (Exists) and universal (Forall) quantifiers, logical disjunction (Or) and conjunction (And) are denoted as $\exists$, $\forall$, $\vee$ and $\wedge$ respectively.
3. NAF(negation as failure) conditions may be denoted as the $\neg$ symbol.
4. Slotted uniterms and frame formulae may be used, as defined by RIF [8].

This RIF-esque language will serve as our "platform independent model", to use an OMG term. That said, we will resort to RIF if, but only if, OWL fails us.

4   The Semantics of Compatibility

In this section, we analyze what it means for things to be compatible, which leads to a discovery whose simplicity somewhat belies its importance. For the sake of generality, we will refer to things in the most abstract sense of that word. To begin with, we explain the notion of relationships in a general way as follows:

Property 1. The ability of a thing to send (receive) messages is denoted as its "produces" ("consumes") property, whose range indicates the type of message that it can handle. In theory, the property's domain, which by default refers to the *owl:Thing* class, could be anything from web services to human beings.

¡owl:Property rdf:ID="produces"¿	¡owl:Property rdf:ID="consumes"¿
¡rdfs:range rdf:resource="&owl;Class"/¿	¡rdfs:range rdf:resource="&owl;Class"/¿
¡/owl:Property¿	¡/owl:Property¿

Next, we define what it means for things that possess the above properties to be compatible with each other. Our interpretations boil down to the inference of the *rdfs:subClassOf* property, which is analogous to the subset operation.

Property 2. For some thing to be compatible with another, the former may only produce messages that are consumable by the latter. That is, the extent of messages produced must be a subset of the extent of messages consumed, or else the producer runs the risk of emitting messages the consumer cannot comprehend.

¡owl:TransitiveProperty rdf:ID="compatibleWith"/¿
$\forall$?p ?c (compatibleWith(?p ?c)   :− $\wedge$ (
produces(?p ?ps) owl : subClassOf(?ps ?cs)) consumes(?c ?cs))

Now, let's define a transitive property *priorVersion(?x ?y)* along the lines of its OWL namesake. This helps in relating things separated by time as shown next.

Property 3. For some newer version of a thing to be backward compatible with an older version, the former must be able to consume all messages that are consumable by the latter. In other words, the extent of the messages consumed by the former must be a superset of the latter's, otherwise the newer version runs the risk of rejecting messages previously allowed. This is the criteria of upgradeability that must be met if newer versions of consumers are to be introduced seamlessly.

$$¡owl: TransitiveProperty\ rdf:ID="backwardCompatibleWith"/¿$$
$$\forall\ ?x\ ?y\ (\!\quad backwardCompatibleWith(?x\ ?y): -\wedge (priorVersion(?y\ ?x)$$
$$conumes(?x\ ?cx)\quad owl: subClassOf(?cy\ ?cx)\qquad consumes(?y\ ?cy))$$

Property 4. Similarly, for some newer version of a thing to be forward compatible with an older version, the latter may only produce messages that are producible by the former. In other words, the extent of the messages produced by the former must be a subset of that of the latter, otherwise the newer producer runs the risk of emitting messages previously not allowed. This is the criteria of extensibility that must be satisfied if older versions of producers are to be future-proof.

$$¡owl: TransitiveProperty\ rdf:ID="forwardCompatibleWith"/¿$$
$$\forall\ ?x\ ?y\ (\quad forwardCompatibleWith(?x\ ?y): -\wedge (priorVersion(?y\ ?x)$$
$$produces(?x\ ?px)\quad owl: subClassOf(?px\ ?py)\qquad produces(?y\ ?py))$$

From the above, one may draw the conclusion that the act of checking to see if two things are compatible essentially boils down to assessing if the classes representing their messages can be related via the *rdfs:subClassOf* property.

5   The Two-Step Semantic Dance

To evaluate compatibility between those message parts that conform to XML Schemas, we require the OWL avatars of the same. To that end, we propose

---

Algorithm 1. Define the extents of OWL types denoting XML model entities.

1. An XML element is mapped to an OWL class. The identity (i.e. rdf:ID) of the class is taken to be the qualified name (i.e. QName) of the element.
2. An XML child element is mapped to an OWL property of the parent element's class. The property's *rdf:ID* is taken to be the child's QName, with its local name capitalized and prefixed with "has" to distinguish it from the name of child's class, which acts as the property's range.
3. An XML attribute is mapped to an OWL property of the class corresponding to its XML element. The property's *rdf:ID* is taken to be the child's QName, with its local name capitalized and prefixed with "has".

Algorithm 2. Declare the values of OWL individuals denoting XML instances.

1. An XML element is mapped to an OWL individual in its corresponding class (see Algorithm 1.) The individual's *rdf:ID*, if not blank, must be globally unique.
2. An XML child element (attribute) is mapped to an OWL property of the individual corresponding to its parent. For every repetition of that particle, the property has a value that refers to the unique OWL individual corresponding to it.

---

a unique two-pass mechanism not unlike the one used in traditional compilers, wherein the syntax is parsed into tokens, which in turn forms the basis for semantic analysis. For us, parsing involves the derivation of crude OWL ontologies from XML Infosets, which facilitates a rigorous analysis of XML concepts. The first step serves to automate the trivial aspects of mapping XML to OWL, which sets the stage for us to deal with its non-trivial aspects wholly in terms of OWL.

5.1   XML Instance Serialization

In the first pass, we serialize the XML document into an OWL ontology, strictly treating the former as being made up of XML InfoSet [9] items. In other words, we consciously avoid interpreting the elements of XML Schema in the document, if any. Specifically, the methodology for serializing XML into OWL is as follows:

In essence, we make an educated guess as to the composition of OWL classes, subject to the particles present in the XML Instance. Note that these classes are merely approximations, and subsets of those entailed by the actual XML types of the XML particles. An XSLT rendition of these rules is available here [10], which may be used to observe the effects of serialization on XML documents.

5.2   XML Schema Ontologizing

Having thus made a foray into OWL, we now redefine the serialized OWL meanings of XML elements and attributes, so that they get aligned with the intended semantics of their XML types and declarations. For completeness' sake, we run the entire gamut of schema components. For most use cases, such as XML validation, a grasp of the core semantics of XML Schemas will suffice. However, since our focus is compatibility, we complement our rule set so as to allow for XML comparison. Further, our algorithm may be recursively applied to the XML Schema for Schemas (XSS). Given that XSS is XML's meta-meta-model, the resulting ontology may potentially serve as an XML Upper Ontology (XUO).

6   The Semantics of XML Schemas

Herein, we analyze the abstract data model of XML Schema in a bid to formalize its meaning. Without loss of generality, we take XSS to be our (logical) starting point, as it is a good representative sample in that it practices what it preaches (for its post-serialization ontology, see [11].) To reiterate, serialization occurs at the XML InfoSet (XIS) level, ergo one cannot expect to gain any insights from the resulting ontology that cannot be inferred by reviewing the XML Instance. Accordingly, for every schema component present, we will redefine its semantics, and assess its compatibility property. We use the namespaces and entities as defined by OWL, except that the default is *xsd*. In addition, we reserve the *xuo* namespace for generic rules. For brevity, we capture the semantic equations separately in [12], and make cross-references to them in the margin notes.

Definition 1. *An XML complex or simple type definition maps to an OWL class, which describes the type's name or, if missing, its anonymous rdf:ID.*

The XML top-level schema element maps to an OWL ontology that describes specifically the target namespace attributed to the schema element.

Property 5. In general, a complex or simple type is compatible with another if the former is an *owl:subClassOf* of the latter.

Definition 2. *We define, as a placeholder forall global elements and attributes, a class using the schema element's rdf:ID, which will act as the schema top entity.*

Property 6. For a schema to be compatible with another, it may only define global elements defined in the latter (consumer-side) schema. However, it may sub-class the types of the consumer-side elements that it contains, if required.

Definition 3. *An XML element or attribute declaration turns into an OWL property. To that end, a property constructor known as* xuo:property *is defined, which capitalizes the XML name and prefixes it with "has", as is standard practice. The property's range denotes the explicitly named or anonymously declared type of the declaration, whereas its domain is the type in which the element was locally declared or, if no such type exists, the schema top's class.*

Property 7. An element (attribute) is said to be compatible with respect to another if it has the same name and it conforms to the same or a compatible complex (simple) type. Additionally, since the types are context-sensitive, the parents of the element (attribute) must also be compatible. As serendipity would have it, the *owl:subPropertyOf* property enables us to check the extents of both the domain and range of the element (attribute) in one shot.

Corollary 1. *The name of the simple or complex type that an element or attribute conforms to is immaterial for computing compatibility.*

Definition 4. *The "minOccurs" ("maxOccurs") attribute of an XML element declaration translates into the "minCardinality" ("maxCardinality") attribute of the corresponding OWL property. If not set, it defaults to 1. The "use" attribute of an XML attribute declaration maps to cardinality constraints on the corresponding OWL property. If its value is "optional" (or "prohibited"), the "minCardinality" ("maxCardinality") is assigned the value 0. Or, if it is missing or has the value "required", then its "minCardinality" ("maxCardinality") is 1.*

Property 8. An element or model group cannot be compatible with another if it extends the consumer's range of allowable occurrences, in either direction.

Corollary 2. *An optional element is not compatible with a mandatory element. However, a mandatory element is compatible with an optional element. An element that is optional may be dropped without adversely affecting compatibility.*

Definition 5. *Annotations appearing inside of an XML type definition, declaration or instance turn into rdfs:comment of the corresponding OWL class, property or individual. In particular, its "documentation" element is captured.*

Property 9. To be compatible, application-centric elements such as "appinfo" present in a consumer-side construct must be retained, if nothing else.

Definition 6. *The XML choice model group naturally turn into the rdf:Alt container, which is a group of alternative resources. Specifically, we construct a class that acts as an rdf:Alt collection, whose members include one property restriction for every element in the choice group. Similarly, the sequence and all groups turn into rdf:Seq and rdf:Bag, which denote ordered and unordered containers resp.*

Property 10. A choice model group is compatible with another if its set of particles is a subset of that of the other, and each of those particles is in turn compatible with the corresponding particle in the other (i.e. consumer-side) group.

Definition 7. *If an element declaration has a true-valued "nillable" attribute, then the type in which it is declared allows that element to appear just to signal its non-existence, in which case it should have a true-valued "xsi:nil" attribute.*

Property 11. A "nillable" element is incompatible with one that is not.

Definition 8. *A simple type having a list type constrains its individuals to take on values that are a space-separated list of atomic pieces of data, each of which in turn is constrained by the declared item type.*

Property 12. A simple type is compatible with another only if the item type of the former (producer) type is a sub class of that of the latter (consumer) type.

Definition 9. *A simple type having a union type constrains its individuals to take on values whose type is one of those declared in its member type set.*

Property 13. A simple type is compatible with another only if the set of member types of the former is a subset of the latter's, with the caveat that a given member type of the former may sub-class the corresponding one of the latter.

Definition 10. *A facet of a simple type imposes a sub-class restriction on it, above and beyond that implied by the type's base class. Without loss of generality, we show here the rule for the "pattern" facet which piggybacks on "fn:match".*

Property 14. A compatible simple type cannot relax facet-defined constraints.

Definition 11. *An XML import element that points to an heterogeneous name space causes the ontology it references to be imported into the current ontology. Similar rules may be defined on the include and redefine elements, although that will require navigating to the external schema location and retrieving its ontology.*

Property 15. An included or redefined schema is equivalent to having it exist inline. Compatible ontologies must import compatible heterogeneous schemas.

Definition 12. *A complex type inherits the particles declared under its "ComplexContent" element, if present. This rules simplifies the act of locating the domains of properties, which can be any kind of particle. A complex type derived by extension from a base type is essentially a union of the base class along with its own restrictions. A simple/complex type derived by restriction from a base type sub-classes the base class, and disallows properties not present in the base.*

Property 16. For two complex types to be compatible, the hierarchy of extensions and restrictions is immaterial as long as the extents satisfy the property of *owl:subClassOf.* In other words, their effective particle sets must be compatible.

Definition 13. *A substitution group's head must be a super property of that of its member's. To see why, observe that the head's type (range) is the base of all member types, plus the member inherits the head's parent type (domain.)*

Property 17. Anelement cannot be compatible with another if the producer's substitution group has elements not present in that of the consumer.

Definition 14. *An "abstract" type implies that its extent is owl:Nothing.*

Property 18. An abstract type is not compatible with one that is not.

Definition 15. *If a complex type has a "final" attribute, its class cannot be derived from, by restriction, extension or both, based on its value. Similarly, if it has a "block" attribute, it can be substituted by either no one or just not by elements that are derivations by restriction or extension, based on its value.*

Property 19. For a type to be compatible with another, the former's effective block (and final) value must be a subset of that of the latter's.

Lemma 1. *First, we define a xuo:classOf property that maps an XPath to a representative owl:Class, made up of nested owl:Restrictions, one for each step.*

Lemma 2. *Second, we define a property to convey the domain (range) of a keybase (i.e. unique, key or keyref element), as a function of its selector (fields.)*

Definition 16. *Every value in an unique element's selected set (domain) maps to a distinct value in its field set (range), which owl:FunctionalProperty enforces.*

Property 20. Anunique key is said to be compatible with another if the former's uniqueness constraint is an *owl:subPropertyOf* of the latter's. In other words, the producer's selection and field set may not surpass those of the consumer.

Definition 17. *We define the xuo:KeyProperty to denote a property that is functional as well as mandatory. It may be likened to the concept of primary keys in databases. Every value in a key element's selected set (domain) must map to a distinct value in its field set (range), which xuo:KeyProperty enforces.*

Property 21. The compatibility rule for keys is similar to the unique one, except that *xuo:KeyProperty* should be used instead of *owl:FunctionalProperty.*

Definition 18. *We use owl:InverseFunctionalProperty to enforce referential integrity from a keyref to its key. In other words, if the keyref property has a value then it implies that the key property must also have a corresponding value.*

Property 22. The compatibility rule for keyrefs is similar to the last one, except that *owl:InverseFunctionalProperty* is used instead of *owl:FunctionalProperty.*

Definition 19. *An "any" ("anyAttribute") inside a complex type denotes a wild-card that is substitutable by any element. The caveat is that if there is a "namespace" attribute specified, then only elements defined in it may be substituted.*

Property 23. A complex type having a wildcard is compatible with another, only if the latter also has that wildcard at the same position. However, the former type may declare a named particle in lieu of a wildcard present in the latter.

Definition 20. *The act of serialization classifies OWL individuals weakly with the XML element's QName. To classify it strongly, we look for a property whose range is the element's type, and domain is the parent element's type. If the property exists, its range serves as the individual's strong type. Note that it is possible for named particles to have different types in different complex contents.*

7   Conclusions and Future Work

Herein, we thoroughly examined various foundational aspects of data compatibility. As long as there are peers exchanging document-style messages, as is the prevalent B2B and SOA use case, our sanity checks have a role. In time, the transition from OWL axioms to FOL rules may become seamless, but we're not there yet. For lack of a standard OWL-aware rule language, we wrote ours in a language-independent way. In the process, we realised that the critical success factors of any such language should include intuitiveness, readability and interoperability. Given that XML is a key building block of the web services stack, the theories fleshed out here may be applicable therein. In particular, an upper ontology for the services space may be gleaned from WSDL schemas using the outlined approach. Our hope is that this paper will serve as groundwork for formalizing the obvious and not-so-obvious meanings of XML-based meta models.

References

[1] Extensible Markup Language (XML) 1.0. W3C Recommendation (August 16, 2006)
[2] XML Schema Part 0: Primer Second Edition. W3C Recommendation (October 28, 2004)
[3] Web Services Description Language Version 2.0 Part 0: Primer (March 2006)
[4] OWL Web Ontology Language Reference. W3C Recommendation (February 10, 2004)
[5] RIF Core Design. W3C Working Draft (March 30, 2007)
[6] SWRL: A Semantic Web Rule Language. W3C Member Submission (May 21, 2004)
[7] Berners-Lee, T.: Experience with N3 rules. W3C Rules Workshop (April 2005)
[8] Core/Slotted Conditions. W3C RIF-WG Wiki (June 28, 2007)
[9] XML Information Set, 2nd (edn.) W3C Recommendation (February 2004)
[10] XML To OWL XSLT. http://geocities.com/karthick_psu/xml-to-owl.xsl
[11] Serialized Schema For Schemas. http://geocities.com/karthick_psu/xss-owl
[12] XML Semantic Rules. http://geocities.com/karthick_psu/xml-rules.pdf

Exploiting E-C-A Rules for Defining and Processing Context-Aware Push Messages

Thomas Beer[1], Jörg Rasinger[1], Wolfram Höpken[1], Matthias Fuchs[1],
and Hannes Werthner[2]

[1] eTourism Competence Center Austria (ECCA), Technikerstrasse 21a
A-6020 Innsbruck, Austria
{firstname.lastname}@etourism-austria.at
[2] Institute for Software Technology and Interactive Systems
Vienna University of Technology, Favoritenstrasse 9-11/188
A-1040 Vienna, Austria
hannes.werthner@ec.tuwien.ac.at

Abstract. The focus of this paper is to show that the E-C-A paradigm offers an excellent approach for specifying the behavior of *context-aware information push services*. Such a service enables its operator to provide the users with tailored messages related to their current situation (context). The paper introduces CAIPS, an implementation of such a service for the tourism domain. The underlying E-C-A rules are presented and the design of the associated rule-engine is described. The engine's rule-interpreter is based on event-notification services and the object-oriented query-language HQL. The paper further presents a graphical high-level editor which supports business-experts in "writing" the CAIPS E-C-A rules. The presented approach enables the rapid development of new tailored messages (related to the user's context) without the need to modify the underlying application, i.e. without the trouble of writing new code for new message types.

Keywords: event-condition-action, reaction rules, context-awareness, push, interpreted rules.

1 Introduction

In general two different information dissemination approaches can be distinguished, namely *push* and *pull* [6, 19]. The traditional pull approach requires that users know a priori where and when to look for information or that they spend an inordinate amount of time polling known sites for updates. Push-based services in turn, relieve their users from these burdens, i.e., the user is not forced to search for relevant information himself and automatically receives the information which is of a high relevance to him, thereby reducing the effort for gathering information [6, 9].

Context-aware services aim at tailoring services to the (user's) context [20]. A *context-aware information push service* (CAIPS) is therefore defined as a service

A. Paschke and Y. Biletskiy (Eds.): RuleML 2007, LNCS 4824, pp. 199–206, 2007.
© Springer-Verlag Berlin Heidelberg 2007

which actively provides its users with tailored information (also denoted as *push-message*) regarding the topical situation (also referred to as *context*). A typical push-message in the tourism domain would be, for example, a message informing tourists about suitable indoor-events in the case of disadvantageous weather changes.

This article proposes an approach for the exploitation of E-C-A rules for specifying the behavior of context-aware information push services. Section two gives an overview of the CAIPS system. The rule-engine's design and its utilized rules are presented in section three and four. Section five presents the rule-editor, which supports domain experts in creating CAIPS E-C-A rules. Conclusions and future work are presented in section six.

2 Context-Aware Information Push Service (CAIPS)

This section introduces CAIPS, an implementation of a context-aware information push service for the tourism-domain [1]. One of the key requirements of the proposed system is "ease of use" for its operator. Put differently, the tourism expert can straightforwardly define *when* to send *what* to *whom*. The proposed approach to address this requirement is to define push-processes through E-C-A *rules*. Rules are a well known technique in the field of knowledge representation [5, 24]. They provide an excellent trade-off between readability of the knowledge representation and formal requirements. Conditional clauses (i.e. "if *condition* then *action*" clauses) were already used in pre-Christian times to express activity-instructions [17]. It therefore can be assumed that rules are well known to prospective users. Using rules addresses one of the prior objectives when designing such a system, namely the similarity to human thinking and the continuous improvement of usability [3].

In CAIPS, the tourism-expert utilizes E-C-A rules to *describe* both, the message's content and the situation, when the message is sent (cf. section 0). Fig. 1 illustrates the message definition and processing of CAIPS. The operator (tourism-expert) creates new message-types using E-C-A rules. He is thereby supported by a high-level graphical editor (cf. section 0), allowing him to easily specify the *message-firing-situation* (cf. section 0) and the message-content. As an integrated part of the etPlanner-framework [16] tourists may subscribe to these message-types within the (mobile) web-application. In the course of the subscription the users must indicate both, information preferences (e.g. sending time, communication channel) and preferences regarding message content (e.g. message should contain a *music-event* recommendation). This information is used for creating personalized content dynamically at the time of the push-rule execution. This approach stands in contrast to conventional *publish-subscribe systems* [22] where the message is produced and the content is matched to the preferences of the potential recipients. Instead, the CAIPS approach produces *tailored messages on demand*.

The message generation and sending process (i.e. the rule interpretation) is controlled by the CAIPS rule-engine, an integral part of the CAIPS system. The next section describes its design in detail.

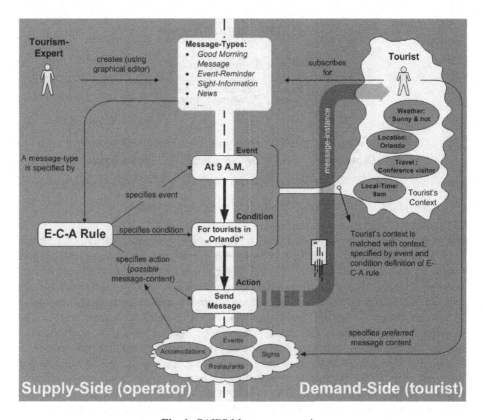

Fig. 1. CAIPS Message processing

3   CAIPS: Rule-Engine

The rule-engine is responsible for interpreting rules and subsequent message-generation. As such, CAIPS is based upon concepts known from knowledge-based systems (KB systems). An important feature all KB systems have in common is the separation between knowledge about the domain of discourse (knowledge-base) and its processing (also referred to as reasoning) [3]. The design of the CAIPS Rule-Engine (RE) is illustrated in Fig. 2.

3.1   Rule-Interpreter

The rule-interpreter is one of the fundamental building-blocks of the rule-engine. It interprets the rules according to the underlying fact-base and thereby controls the application's behavior. It is composed of three sub-components: the *Event-Handler*, the *Condition-Evaluator*, and the *Action-Manager*.

The event handling process is realized similar to a (distributed) event-notification service (ENS) [14]. Such a service informs entities (subscribers) about *events* they are interested in. Similar to [14] an *event* is defined as a state transition of an information

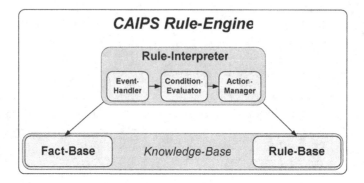

Fig. 2. Rule-Engine's Core Components

object at a certain point of time, e.g. a user approaches a certain location, or a certain point of time is reached. An ENS is composed of event-triggers (services which report an event) and an event-handler. CAIPS distinguishes *external* and *internal* event-triggers. External events are triggered by external services, such as a weather change which could be reported by a weather server. Internal events are triggered "internally" by a state change within the fact-base, such as a change to the user's profile. The message-based (asynchronous) communication between event-trigger and –handler enables the decoupling of event-generation and the subsequent handling-process.

The *Event-Handler* determines the rule(s) which are configured to react to the incoming event. This is done by a matching algorithm which "compares" incoming event-messages with the rule's event-profile. This event-profile (which is formalized based on an event language similar to [14, 21]) denotes the rule's interest in the incoming event (cf. section 0).

After identifying the "matching" rules, the condition evaluation process is triggered. The CAIPS condition-evaluator transforms the rule's condition definition (cf. section 0) into the appropriate query-language and delegates the query processing to the related query engine. Clearly, the utilized query language must be compliant to the fact-base's implementation (cf. section 0). CAIPS uses the object-oriented query-language HQL [13] to access the object-oriented fact-base. The query-results, comprising a number of potential receivers, are then forwarded to the action-manager.

The action-manager determines the appropriate action-processor according to the rule's action definition and subsequently triggers the action-execution. The action-processor completes the remaining tasks, i.e. it creates and sends the personalized messages [1].

3.2 Knowledge Base

The knowledge-base stores the "behavioral" knowledge of CAIPS, its *static* (i.e. the facts) and its *procedural* knowledge (i.e. the rules) [18]. It can be separated into two further sub-entities, namely the *fact-base* and the *domain & general knowledge* part (i.e. the rule-base). The fact-base contains problem specific knowledge, i.e., it represents the current state of the problem domain (i.e. the user's context). The domain & general knowledge part stores domain specific and general knowledge about objects and their relationships, and heuristic knowledge which constitutes the "art of good guessing" in the domain (also known as rules-of-thumb or good judgment)[9]. In CAIPS the heuristic is constituted by the E-C-A rules.

The fact-base is based on object-oriented and domain modeling [8] design techniques. The facts (i.e. the user's context) are represented by objects which are stored using an object/relational mapper [12]. The acquisition of coarse grained context data from the various sensors available in the application domain and its translation into objects is realized by appropriate context integration- and preprocessing-services [1].

The fact-base provides information for specifying the *message-firing-situation (MFS)*. The MFS occurs when the user's context state (his "real-world situation") matches a predefined state, specified by the rule's event- and condition statement.

The rule-base is the central repository for maintaining the rules. It provides functionality for creating, updating, deleting, and persisting CAIPS E-C-A rules. Currently the rule-base is implemented using a XML-based repository.

4 CAIPS Rule Markup Language

The utilized rule language is composed of query- and event-languages similar to the language presented in [21]. An example rule is presented at the end of this section. Each rule consists of three sections, the *event*, the *condition*, and the *action* definition.

Event

CAIPS utilizes an event language similar to the proposed language in [14]. This language enables the specification of primitive and composite events. Compared to [23], CAIPS supports the specification of *passive* events only; active events are covered by utilizing internal event-triggers (cf. section 0) which are "configured" outside of the rule-definition (cf. section 0). The event section specifies the event-profile (compare *client-profile* in [14]) which is utilized for filtering all rules which match an "incoming" event. The profile consists of a set of key-value pairs and an event type which categorizes the occurred event. An example XML representation of such an event-profile specification is shown in lines 2 to 8.

Condition

The condition definition is used to determine the set of users for which the occurred event is of high relevance. Restrictions are specified to narrow down the potential set of receivers (i.e., the users who will obtain generated push-messages). An example of a trivial condition statement is shown in lines 9 to 14.

Action

The action-definition specifies the action to be executed. The action, shown in lines 15 to 41, shows the definition of a "sendMessageAction". The embedded elements are used to configure the action-implementation. More precise, the parameters specify information to be passed onto the message generator.

The following example illustrates the representation of an E-C-A rule in CAIPS:

```
1:    <rule active="true" id="397705181623228736">
2:        <event id="3977051816232287345">
3:            <type>TimeEvent</type>
4:            <param>
5:                <key>cron</key>
6:                <value>20 45 07 08 2007</value>
7:            </param>
8:        </event>
9:        <condition id="345789152443985792">
10:           <conditionType>HQL</conditionType>
11:           <conditionstatement>from User as u where
12:               u.prefs.identifier='GoodMorningMessage'
13:           </conditionstatement>
14:       </condition>
15:       <action id="98101636422723184" type="SendMessageAction">
16:           <content>
17:               <recommendation>
18:                   <type>Event</type>
19:                   <parameter>
20:                       <customerProperty>
21:                           <attributeName>startdatefrom</attributeName>
22:                           <attributeValue>&system.RuleExecutionDate</attributeValue>
23:                       </customerProperty>
24:                       <customerProperty>
25:                           <attributeName>classificationvalue</attributeName>
26:                           <attributeValue>&user.sight.classification</attributeValue>
27:                       </customerProperty>
28:                       <customerProperty>
29:                           <attributeName>city</attributeName>
30:                           <attributeValue>&user.address.city </attributeValue>
31:                       </customerProperty>
32:                       <customerProperty>
33:                           <attributeName>startdateto</attributeName>
34:                           <attributeValue>&system.RuleExecutionDate</attributeValue>
35:                       </customerProperty>
36:                   </parameter>
37:               </recommendation>
38:               <template>
39:                   <templateidentifier>GoodMorningMessage</templateidentifier>
40:               </template>
41:           </content>
42:       </action>
43:   </rule>
```

5   Rule Wizard

In section 0 the importance of supporting domain-experts in creating the CAIPS E-C-A rules was emphasized. CAIPS addresses this requirement by a graphical editor offering a high-level interface for creating CAIPS rules. The editor provides wizards for each part (event, condition, action) of a CAIPS rule, i.e. the business-expert can create a rule by a few "clicks" only. As an example, the event profile is built by selecting an "event" from a drop-down list. Further, the definition of condition-statements is facilitated by using forms. The approach is comparable to the "query by forms" technique presented in [11]. The screenshot in Fig. 3 illustrates its usage

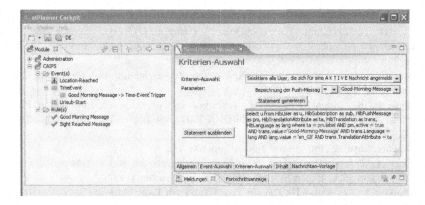

Fig. 3. CAIPS rule-editor, creating a rule's condition definition

within the CAIPS rule-editor. The underlying HQL statement, which represents the condition definition (cf. sections 0 and 0), is generated according to the chosen form fields. Expert users may edited the condition definition directly.

The editor also offers a wizard for the creation of internal event-triggers (cf. section 0). A prominent example is the creation of a time-event-trigger, which may be used to define events based on a time-schedule (comparable to cron-triggers).

6 Conclusion and Future Work

The paper demonstrated that interpreted E-C-A rules provide an excellent approach for controlling the behavior of context-aware push applications. It presented their applicability by using such rules in CAIPS, a prototypical implementation of a context-aware push service for the destination of Innsbruck. First field trials indicate high potential of the proposed service [1]. Based on these field trials further implementations for *Dolomiti-Superski* [7] (one of the largest skiing destination in Europe) and *Hofer-Reisen* [15] will take place until the end of 2007.

Future research will include an evaluation how the proprietary CAIPS rule-language may be complemented or even substituted by *Reaction RuleML* [23]. Furthermore it will be evaluated how the design of the fact-base effects tourism-experts in creating a rule's condition section, e.g. using conceptual queries [4] instead of query by forms.

References

1. Beer, T., Höpken, W.R., Zanker, M., Rasinger, J., Jessenitschnig, M., Fuchs, M., Werthner, H.: An Intelligent Information Push Service for the Tourist Domain. In: European Conference on Artificial Intelligence (Workshop on Recommender Systems), Riva del Garda (2006)
2. Beer, T., Fuchs, M., Höpken, W.R., Rasinger, J., Werthner, H.: CAIPS: A Context-Aware Information Push Service in Tourism. In: Sigala, M., Mich, L., Murphy, J. (eds.) Information and Communication Technologies in Tourism 2007, Wien, pp. 129–140. Springer, Heidelberg (2007)

3. Beierle, C., Kern-Isberner, G.: Methoden wissensbasierter Systeme. Hagen, Vieweg (2003)
4. Bloesch, A.C., Halpin, T.A.: ConQuer: A Conceptual Query Language. In: 15. Int. Conference on conceptual modeling, Cottbus, Germany. LNCS, Springer, Heidelberg (1996)
5. Brachman, R.J., Levesque, H.J.: Knowledge Representation and Reasoning. Morgan Kaufmann, San Francisco (2004)
6. Cheverst, K., Mitchell, K., et al.: Investigating Context-aware Informaiton Push vs. Information Pull to Tourists. In: MobileHCI 2001 workshop on HCI with Mobile Devices, Lille, France (2001)
7. Dolomiti Superski, http://www.dolomitisuperski.com
8. Evans, E.: Domain-Driven Design. Addison-Wesley, Boston, MA (2004)
9. Feigenbaum, E.A.: Expert Systems: Principles and Practice (1992)
10. Franklin, M.J., Zdoni, S.B.: Data In Your Face: Push Technology in Perspective. In: ACM SIGMOD International Conference on Management of Data, ACM Press, Seattle, Washington, USA (1998)
11. Halpin, T.: Information Modeling and Relational Databases - From Conceptual Analysis to Logical Design. Morgan Kaufmann, San Francisco (2001)
12. Hibernate O/R Mapper, http://www.hibernate.org/
13. Hibernate Query Lang. http:// www.hibernate.org/ hib_docs/ reference/ en/html/ queryhql. html
14. Hinze, A.: A-MEDIAS: Concept and Design of an Adaptive Integrating Event Notification Service. Freie Universität. Berlin, Mathematik u. Informatik. PhD thesis (2003)
15. Hofer Reisen, http://www.hofer-reisen.at
16. Höpken, W., Fuchs, M., Zanker, M., Beer, T., Eybl, A., Flores, S., Gordea, S., Jessenitschnig, M., Kerner, T., Linke, D., Rasinger, J., Schnabl, M.: etPlanner: An IT framework for comprehensive and integrative travel guidance. In: Hitz, M., Sigala, M., Murphy, J. (eds.) Information and Communication Technologies in Tourism 2006, Wien, pp. 125–134. Springer, Heidelberg (2006)
17. Jaynes, J.: The Origins of Consciousness in the Breakdown of the Bicameral Mind. Princeton University Press, Princeton (1976)
18. Karagiannis, D., Telesko, R.: Wissensmanagement. Konzepte der Künstlichen Intelligenz und des Softcomputing: Oldenbourg (2001)
19. Kendall, J.E., Kendall, K.E.: Information Delivery Systems: An Exploration of Web Pull and Push Technologies. Communications of the AIS 1(4) (1999)
20. Kranenburg, H.: A context management framework for supporting context-aware distributed applications. Communications Magazine 44(8), 67–74 (2006)
21. May, W., Alferes, J.J., Ricardo, A.: Active Rules in the Semantic Web: Dealing with Language Heterogeneity. In: Adi, A., Stoutenburg, S., Tabet, S. (eds.) RuleML 2005. LNCS, vol. 3791, pp. 30–44. Springer, Heidelberg (2005)
22. Mühl, G.: Large-Scale Content-Based publish subscribe systems. Informatik. Darmstadt, Technische Universität. PhD thesis (2002)
23. Reaction RuleML, http://ibis.in.tum.de/research/ReactionRuleML/
24. Sowa, J.F.: Knowledge Representation: Logical, Philosophical, and Computational Foundations. Brooks Cole Publishing, Pacific Grove, CA (2000)

The Use of Ontologies and Rules to Assist in Academic Advising

Girish R. Ranganathan and J. Anthony Brown

University of New Brunswick, 15 Dineen Dr., Fredericton, New Brunswick, E3B 5A3, Canada
{girish.ranganathan,Anthony.brown}@unb.ca

Abstract. Undergraduate academic advisors, schools of graduate studies, and other organizations are often responsible for understanding and evaluating high-school and university transcripts from culturally and organizationally different institutions. Such a task raises several problems including diverse grading approaches, incompatible academic credit systems, and translating between languages. Although many such problems are most sensitive in a world (or international) evaluation, they can occur between institutions within the same country as well. The present paper proposes a possible methodology toward overcoming such conflicts using a University-Course-Credit-Grade (UCCG) ontology and translation rules for interoperability of credit systems and grade systems between institutions. As proof of concept, the UCCG ontology is initially populated with example instances from four institutions, using Protégé-2000 to build the ontology, RDF(S) to store it, and POSL to store the university instances and translation rules.

1 Introduction

The globalization of our culturally and organizationally diverse society and the popularization of student exchanges and international study have created new challenges and opportunities for e-Learning applications. For example, it is becoming increasingly common for students to transfer course credits or degrees between institutions at an international level. In such a scenario, undergraduate academic advisors, schools of graduate studies, and other organizations are often required to understand and precisely evaluate high-school and university transcripts. Conducting a comprehensive evaluation of a foreign transcript must take into account many important issues such as diverse grading approaches, incompatible academic credit systems, and translating between languages. For example, many universities in North America assign letter grades from "F" to "A+" (0.0 to 4.3) or percentage grades from "0%" to "100%"; however, in many East European universities, a 5-point ("2" to "5", with "2" being the lowest) grading scheme is applied. Furthermore, in Germany, a similar 5-point scheme is used, but the German system ranges from "1" to "5" with "5" being the lowest. These problems are further compounded by inconsistent passing grades in different systems; for example, a "D" may be considered a pass in the North American system, but directly translates to a 2.0 within the 5-point system, which is considered a failing grade. In many cases, a direct conversion is not reasonable and equivalent ranges must be found; for example, a 4.0 within the 5-point system could

A. Paschke and Y. Biletskiy (Eds.): RuleML 2007, LNCS 4824, pp. 207–214, 2007.

be equivalent to any value within a range of 2.7 and 3.3 in the 4.3-point system. While these problems are most sensitive in international evaluation, they can even appear within two institutions in the same country. For example, the University of New Brunswick uses a "credit hour" system based on the number of weekly hours required by a course. McMaster University, however, assigns three "credits" for every single-term course regardless of the time spent per week. These few examples demonstrate the many difficulties encountered when attempting to achieve interoperability between transcripts from different institutions and identify the need for an automatic e-Learning application to facilitate the task. This paper presents the first steps toward overcoming these issues using an ontology and rule-based conversion of credits and grades, with the intention of delivering a transcript issued by one institution to the context of another. The presented work can be used as an integral component in areas such as learning object interoperability [1-3], adaptive learning [4], [5], and personalization [6], [7] on the Semantic Web infrastructure [8], [9].

2   The University-Course-Credit-Grade Ontology

The central components of our approach are a University-Course-Credit-Grade (UCCG) ontology and a set of rules for translation between inconsistent credit and grade systems. A lightweight ontology [10] was developed using Protégé-2000 and stored in the Resource Description Framework Schema (RDFS) [11] syntax:

```
<?xml version='1.0' encoding='UTF-8'?>
<!DOCTYPE rdf:RDF [
     <!ENTITY rdf 'http://www.w3.org/1999/02/22-rdf-syntax-ns#'>
     <!ENTITY kb 'http://protege.stanford.edu/kb#'>
     <!ENTITY rdfs 'http://www.w3.org/2000/01/rdf-schema#'>
]>
<rdf:RDF xmlns:rdf="&rdf;"
     xmlns:kb="&kb;"
     xmlns:rdfs="&rdfs;">
<rdfs:Class rdf:about="&rdfs;Resource"
     rdfs:label="Resource">
</rdfs:Class>
<rdfs:Class rdf:about="&kb;Credit"
     rdfs:label="Credit">
     <rdfs:subClassOf rdf:resource="&rdfs;Resource"/>
</rdfs:Class>
<rdfs:Class rdf:about="&kb;University"
     rdfs:label="University">
     <rdfs:subClassOf rdf:resource="&rdfs;Resource"/>
</rdfs:Class>
<rdfs:Class rdf:about="&kb;course"
     rdfs:label="course">
     <rdfs:subClassOf rdf:resource="&rdfs;Resource"/>
</rdfs:Class>
<rdf:Property rdf:about="&kb;credit_type"
     rdfs:label="credit_type">
     <rdfs:domain rdf:resource="&kb;Credit"/>
     <rdfs:range rdf:resource="&rdfs;Literal"/>
</rdf:Property>
...
</rdf:RDF>
```

Positional Slotted Language (POSL) syntax [12] was used to implement the ontology instances (and rules and facts – Section 3). It should be noted that POSL was used primarily for its readability and ease of use and ontology instances written in POSL can be directly translated to RDF for storage and interoperability. The UCCG ontology was initially populated with instances from four universities and can easily be further populated with new instances corresponding to additional universities.

The ontology consists of four entities: "University", "Course", "Credit", and "Grade". "University" has the attribute "name" which accepts values of type "string":

```
name(univ_unb : University, "UNBF" : String).
```

where,   name – is the attribute;
          univ_unb – is an instance of the class "University";
          UNBF – is the value of type string for the attribute;

The class "Credit" has two attributes: "credit_type", which defines the credit scheme (i.e. credit hour system) and "credit_value", which defines a particular value within the chosen credit scheme. For example, a 4.0 credit hour course at the University of New Brunswick would define "credit_unb" as an instance of the class "Credit" and use the "c-hour" system with a corresponding value of 4.0, as follows:

```
credit_type(credit_unb : Credit, "c-hour" : String).
credit_value(credit_unb : Credit, 4.0 : Real).
```

In a similar fashion the "Grade" class has two attributes: grade_type, which defines the grading scheme (e.g. 4.3-point system) and "grade_value", which defines a particular value within the chosen grading scheme, as follows:

```
grade_type(grade_unb : Grade, 4.3 : Real).
grade_value(grade_unb : Grade, 4.0 : Real).
```

The primary concept in the UCCG ontology is the "Course" class, which has four attributes: name, credit_scheme, belongs_to and grading, shown below for a particular instance:

```
name(course_unb : course, "VLSI" : String).
credit_scheme(course_unb : course, credit_unb : Credit).
belongs_to(course_unb : course, univ_unb : University).
grading(course_unb : course, grade_unb : Grade).
```

3   Conversion Facts and Rules

Extensible sets of rules and facts were manually written, in POSL, for the four universities populating the ontology. The POSL syntax allows rules and facts to be easily translated to Rule Markup Language (RuleML) [13] meaning the ontology can be stored in a platform-independent format for processing, subsequent reuse, integration with other ontologies, and mapping to existing ontology languages and standards.

It should be noted that the presented facts and rules describe conversions that are not symmetric and may not be universally applicable. Instead, the presented conversion facts are the result of an agreement between domain experts (i.e. academic advisors) and must be maintained by the institutions that wish to use them. Because the facts are platform-independent, they can be easily exchanged, altered, or replaced.

There are four groups of conversion facts and rules in the system:

- Grade conversion facts, which relate grades between two grading systems;
- Credit conversion facts, which relate credit values between two university systems;
- Grade conversion rules, which represent conversion between grading schemes;
- Credit conversion rules, which represent conversion between credit schemes;

Example grade conversion facts representing conversions from the 4.3-point system (referred to as basic) to a percentage system are given below:

```
Grade_basicTopercent(?basic : Real, 95.0 : Real) :-
greaterThan(?basic, 4.0 : Real), lessThan(?basic, 4.4 : Real).
Grade_basicTopercent(?basic : Real, 90.0 : Real) :-
greaterThan(?basic, 3.7 : Real), lessThan(?basic, 4.1 : Real).
...
Grade_basicTopercent(?basic : Real, 95.0 : Real) :-
greaterThan(?basic, 4.0 : Real), lessThan(?basic, 4.4 : Real).
Grade_basicTopercent(?basic : Real, 90.0 : Real) :-
greaterThan(?basic, 3.7 : Real), lessThan(?basic, 4.1 : Real).
```

Example credit conversion facts representing conversions between the "credit hour" and "academic hour" systems are given below (where 1 credit_hour= 16×academic_hour):

```
Credit_chourToahour(?chour : Real, ?ahour: Real) :-
multiply(?ahour : Real, 16.0 : Real, ?chour : Real).
Credit_ahourTochour(?ahour : Real, 3.0 : Real) :-
greaterThan(?ahour, 40.0 : Real), lessThan(?ahour, 55.0 : Real).
Credit_ahourTochour(?ahour : Real, 4.0 : Real) :-
greaterThan(?ahour, 54.0 : Real), lessThan(?ahour, 73.0 : Real).
```

Example "Credit_convert" rules are shown below. These rules translate a "credit_value" from one "credit_type" to another (e.g. 16 academic hours to 1 credit hour):

```
Credit_convert(?credit_val : Real, ?credit_type : String, ?credit_val :
Real, ?new_credit_type : String) :- equal(?credit_type : String,
?new_credit_type : String).
...
Credit_convert(?credit_val : Real, ?credit_type : String,
?new_credit_val : Real, ?new_credit_type : String) :- equal(?credit_type
: String, "a-hour" : String), equal(?new_credit_type : String, "c-hour"
: String),

Credit_ahourTochour(?credit_val : Real, ?new_credit_val : Real).
```

Example "Credit_converter" rules are shown below. These rules make use of the appropriate lower level "Credit_convert" rule to translate between two Credit systems (e.g. an instance of Credit of type academic hours to credit hours):

```
Credit_converter(?old_credit : Credit, ?new_credit_type : String,
?new_credit_val : Real) :-
credit_type(?old_credit : Credit, ?cgggg3 : String),
credit_value(?old_credit : Credit, ?cgggg4 : Real),
Credit_convert(?cgggg4 : Real, ?cgggg3 : String, ?new_credit_val : Real,
?new_credit_type : String).
```

Example "Grade_convert" rules are shown below. These rules translate a grade from one "grade_type" to another (e.g. 4.0 on a 4.3 system to 85% on a percentage system):

```
Grade_convert(?grade_old : Real, ?grade_old_type : Real, ?grade_new :
Real, ?grade_new_type : Real) :- equal(?grade_old_type : Real, 4.3 :
Real), equal(?grade_new_type : Real, 100.0 : Real),
Grade_basicTopercent(?grade_old : Real, ?grade_new : Real).

Grade_convert(?grade_old : Real, ?grade_old_type : Real, ?grade_new :
Real, ?grade_new_type : Real) :- equal(?grade_old_type : Real, 4.3 :
Real), equal(?grade_new_type : Real, 5.0 : Real),
Grade_basicTofive(?grade_old : Real, ?grade_new : Real).
   ...
Grade_convert(?grade_old : Real, ?grade_old_type : Real, ?grade_new :
Real, ?grade_new_type : Real) :- equal(?grade_old_type : Real, 100.0 :
Real), equal(?grade_new_type : Real, 5.0 : Real),
Grade_percentTofive(?grade_old : Real, ?grade_new : Real).
```

Example "Grade_converter" rules are shown below. These rules make use of the appropriate lower level "Grade_convert" rule to translate between two Grade systems (e.g. an instance of Grade of type 4.3 system to a percent system):

```
Grade_converter(?grade : Grade, ?new_grade_type : Real, ?new_grade_value
: Real) :-
grade_type(?grade : Grade, ?grade_type : Real), grade_value(?grade :
Grade, ?grade_value : Real), Grade_convert(?grade_value : Real,
?grade_type : Real, ?new_grade_value : Real, ?new_grade_type : Real).
```

Within the proposed methodology, if the existing knowledge base contains N institutions (with each using a unique credit and grading scheme), adding a new institution requires the development of N new credit and grade system translators to achieve complete interoperability. As a result, the knowledge base could easily become very large and inefficient as N×(N-1) translators are necessary to achieve interoperability between N unique systems. It would therefore be desirable to implement single, standard interchange systems for all credit and grading schemes. Another (partial) solution involves the use of transitive rules; for example, if there is no existing translation between the "grade_old" and "grade_new" grade systems but the knowledge base has translators from "grade_old" to "temp_type" systems and from "temp_type" to "grade_new" systems, the rules could be transitively applied, as follows:

```
Grade_convert(?grade_old : Real, ?grade_old_type : Real, ?grade_new :
Real, ?grade_new_type : Real) :- Grade_convert(?grade_old : Real,
?grade_old_type : Real, ?temp : Real, ?temp_type : Real),
Grade_convert(?temp : Real, ?temp_type : Real, ?grade_new : Real,
?grade_new_type : Real).
```

The following set of rules is most applicable for academic advising. The rules shown bellow convert the credit value and grade value of an instance of a course from one system to another system, based on the corresponding universities:

```
Course_converter(?course : course, ?univ : University, ?new_credit :
Real, ?new_grade : Real) :-
credit_scheme(?course : course, ?xx : Credit), grading(?course : course,
?xy : Grade), belongs_to(?y2 : course, ?univ : University),
credit_scheme(?y2 : course, ?y3 : Credit), credit_type(?y3 : Credit, ?y4
: String), grading(?y2 : course, ?y5 : Grade), grade_type(?y5 : Grade,
?y6 : Real), Grade_converter(?xy : Grade, ?y6 : Real, ?new_grade :
Real), Credit_converter(?xx : Credit, ?y4 : String, ?new_credit : Real).
```

4 Querying the Ontology

To verify this implementation, the ontology and rules were queried using a University
of New Brunswick course (course_unb – described in   Section 2) to find the
equivalent credits and grades from the University of Prince Edward Island (univ_pei).
The results of query shown below, issued using the OO jDREW Top-Down reasoning
engine [14], are shown in Fig. 1:

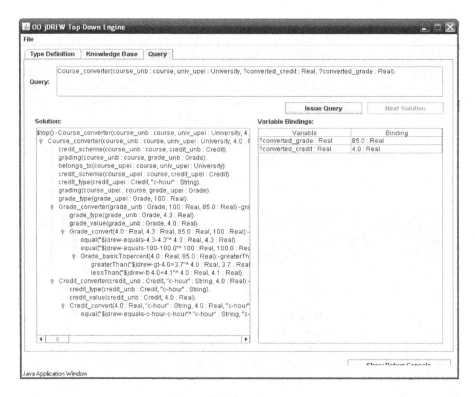

Fig. 1. OOjDrew Snapshot

```
Course_converter(course_unb  :   course,   univ_upei   :   University,
?converted_course : Real, ?converted_grade : Real).
```

The query asks "what is the equivalent credit and grade for the course couse_unb at
the university univ_upei?" The instantiated (fully-expanded) proof tree on the left side

of Figure 1 and the solution variable binding on the right side indicate that the variable values were inferred are as expected:

```
?converted_credit = 4.0 of type Real

?converted_grade = 85.0 of type Real
```

5 Conclusion

This paper proposed a methodology to partially automate the process of understanding and evaluating transcripts from institutions using foreign grading or credit systems during academic advising. The described system uses an extensible UCCG ontology and translation rules to convert credits and grades from a transcript issued by one institution to the context of another. As proof of concept, the UCCG ontology was developed using Protégé-2000 and stored in RDFS syntax, POSL was used to populate the ontology with instances and implement the rules and facts, and OOjDrew was used to issue queries to the populated ontology. It should be noted that while the facts presented here were manually created, it is possible to (semi)automatically generate the facts by parsing and analyzing transcripts and converting the extracted information to POSL syntax. Similarly, conversion facts could easily be prepared by a domain expert or academic advisor in a simple Excel decision table, then parsed into the knowledge base. The work presented here could serve as a key component in areas such as learning object interoperability, adaptive learning, and personalization on the Semantic Web infrastructure.

References

[1] Friesen, N.: Interoperability and Learning Objects: An Overview of E-Learning Standardization. Interdisciplinary Journal of Knowledge and Learning Objects 1, 23–31 (2005)

[2] Biletskiy, Y., Vorochek, O., Medovoy, A.: Building Ontologies for Interoperability among Learning Objects and Learners. In: Orchard, B., Yang, C., Ali, M. (eds.) IEA/AIE 2004. LNCS (LNAI), vol. 3029, pp. 977–986. Springer, Heidelberg (2004)

[3] Biletskiy Y., Boley H., Zhu, L.:A RuleML-Based Ontology for Interoperation between Learning Objects and Learners. UCFV Research Review, Issue 1 (July 2, 2007), (2006) Available http://journals.ucfv.ca/ojs/rr/

[4] Denaux, R., Dimitrova, V., Aroyo, L.: Integrating Open User Modeling and Learning Content Management for the Semantic Web. In: Ardissono, L., Brna, P., Mitrović, A. (eds.) UM 2005. LNCS (LNAI), vol. 3538, pp. 9–18. Springer, Heidelberg (2005)

[5] Aroyo, L., Dolog, P., Houben, G.J., Kravcik, M., Naeve, A., Nilsson, M., Wild, F.: Interoperabilityin Personalized Adaptive Learning. Educational Technology & Society 9(2), 4–18 (2006)

[6] Keleberda, I., Lesna, N., Makovetskiy, S., Terziyan, V.: Personalized Distance Learning based on Multiagent Ontological System. In: Kinshuk, E. (ed.) Proc. of the 4-th IEEE Int. Conf. on Advanced Learning Technologies, IEEE Comp, pp. 777–779. Soc. Press, Joensuu (Finland) (2004)

[7] Keleberda, I., Repka, V., Biletskiy, Y.: Building Learner's Ontologies to Assist Personalized Search of Learning Objects. In: Keleberda, I., Repka, V., Biletskiy, Y. (eds.) ICEC 2006. The 8th International Conference on Electronic Commerce, pp. 569–574. Fredericton, Canada (2006)

[8] Berners-Lee, T., Hendler, J., Lassila, O.: The Semantic Web. Scientific American 284(5), 34–43 (2001)

[9] Oberle, D., Staab, S., Studer, R.: Supporting application development in the semantic web. ACM Transaction on Internet Technology (TOIT) 5/2, 329–358 (2005)

[10] Gomez-Perez, A., Corcho, O., Fernandez-Lopez, M.: Ontological Engineering: with examples from the areas of Knowledge Management, e-Commerce and the Semantic Web (Advanced Information and Knowledge Processing), p. 415. Springer, Heidelberg (2004)

[11] RDF Vocabulary Description Language 1.0: RDF Schema (July 2, 2007) Available http://www.w3.org/TR/rdf-schema/

[12] Boley, H.: POSL: An Integrated Positional-Slotted Language for Semantic Web Knowledge (July 2, 2007) (2004) Available http://www.ruleml.org/submission/ruleml-shortation.html

[13] Boley, H.: The RuleML Family of Web Rule Languages. In: Alferes, J.J., Bailey, J., May, W., Schwertel, U. (eds.) PPSWR 2006. LNCS, vol. 4187, pp. 1–15. Springer, Heidelberg (2006)

[14] Ball, M., Boley, H., Hirtle, D., Mei, J., Spencer, B.: The OO jDREW Reference Implementation of RuleML. In: Adi, A., Stoutenburg, S., Tabet, S. (eds.) RuleML 2005. LNCS, vol. 3791, pp. 218–223. Springer, Heidelberg (2005)

Towards Knowledge Extraction from Weblogs and Rule-Based Semantic Querying

Xi Bai[1,2], Jigui Sun[1,2], Haiyan Che[1,2], and Jin Wang[3]

[1] College of Computer Science and Technology, Jilin University,
Changchun 130012, China
[2] Key Laboratory of Symbolic Computation and Knowledge Engineering of Ministry
of Education, Changchun 130012, China
[3] Institute of Network and Information Security, Shandong University,
Jinan 250100, China
xibai@email.jlu.edu.cn

Abstract. Weblogs (blogs) becomes a very popular medium for exchanging information, opinions and experiences nowadays. However, since new blog pages are constantly issued, finding out helpful information from them becomes a tedious and time consuming work. This paper proposes a system for extracting knowledge hidden in blog pages in Chinese. Before extraction, blog pages are clustered into categories. Then for each category, the knowledge can be extracted based on domain ontologies. Using restrained natural language processing, user can query the KB and the helpful knowledge will be returned based on reasoning about the individuals. KEROB, a prototype of our system, is designed and implemented to fulfill the above functions. The experimental results indicate the superiority of our system.

1 Introduction

More and more people, organizations and companies begin to put information on weblogs (blogs) nowadays. With the stepwise development of blogging techniques, finding out approaches for extracting information from numerous blog pages becomes one of the heated topics. Information extraction (IE) techniques are capable of decoding targeted subject information in documents, and reducing text data into a set of structured core information [1]. However, since the data extracted by traditional IE techniques are not machine-understandable, it can not be processed by computers directly. Moreover, it is not convenient for users to query the data and sometimes the extracted data are deficient. For instance, users may not only search for the current condition of a specific stock but also expect to know its previous condition or developing trend in the future. Traditional IE techniques are helpless in this case.

Motivated by the above analysis, we propose a novel knowledge extraction framework for mining blog. Based on this framework, KEROB, a prototype of our system is designed and implemented, which can automatically extract the machine-understandable knowledge from blog pages in Chinese. Since the

A. Paschke and Y. Biletskiy (Eds.): RuleML 2007, LNCS 4824, pp. 215–223, 2007.

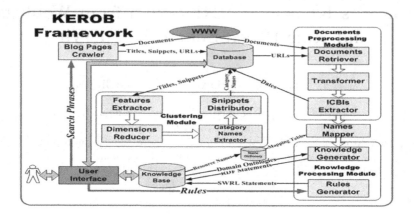

Fig. 1. Framework for KEROB

pre-clustering is employed, the performance of our system is further improved. Moreover, common sense rules and personalized rules can be created according to the users' specific needs. Users can do queries based on restrained natural language processing (NLP) and get the interesting knowledge stored in original blog pages or inferred by the reasoner of KEROB. We sketch the framework for KEROB in Figure 1 and following sections will give the further explanations.

2   Related Work

To the best of our knowledge, among the published works little is dealing with extracting knowledge from blogs. However, some blog mining and blog searching techniques have been proposed recently. Some of them focus on aggregating the entries of the blog pages. Nanno et al. proposed a system that automatically collects and monitors Japanese blog entries, and returns aggregated results [2]. Kurashima et al. proposed a system that can mine association rules between locations, time periods and types of experiences out of blog entries [3]. Moreover, some techniques have appeared to extract personalized information from blog pages. Based on *burst extraction*, Kleinberg et al. proposed *blogWatcher*, a system that can extract hot topics attracting a lot of attentions [4]. Nakatsuji et al. proposed an information filtering approach to automatically transcribe users' interests to a user profile in detail based on service domain ontologies [5]. The above works motivate us to propose an approach for extracting knowledge from blogs and provide users with a convenient interface to create rules and do queries.

3   Snippets Clustering and ICBIs Extraction

In this section, we introduce our approach for clustering snippets and extracting interesting blocks. We first invoke Google AJAX Search APIs[1] to acquire blog

---

[1] http://code.google.com/apis/ajaxsearch

snippets according to the search phrases users input. Then we can get the titles, the content and the original URLs of snippets, which will all be stored into the database. The above task is fulfilled by *Blog Pages Crawler* block in Figure 1.

3.1 Features Extraction and Dimensions Reduction

We use Vector Space Model (VSM) to model the snippets. The feature vector can be denoted by $(tf_{w_1} \cdot idf_{w_1}, tf_{w_2} \cdot idf_{w_2}, ..., tf_{w_n} \cdot idf_{w_n})$, where tf_{w_i} denotes the frequency of the ith word and idf_{w_i} denotes the weighting factors of the ith word. Assume there are m snippets in a corpus which contains total n different terms. Then this corpus can be denoted by a $m \times n$ matrix M. We can use suffix arrays [6] to calculate the frequency of each term. Then we represent a corpus by a term-snippet matrix. *Term* denotes a specific Chinese word and can be extracted based on Chinese-word-segmentation techniques [7]. The dimensions of the feature vectors extracted from Chinese documents are usually too large (maybe larger than 50000). Here, we use Latent Semantic Indexing (LSI) to reduce the dimensions. Firstly, Singular Value Decomposition (SVD) is applied to matrix M and we get $M = USV^T$, where U and V denote *left singular vectors* and *right singular vectors* of M respectively, and S denotes a matrix having the singular values of M ordered decreasingly along its diagonal. After the selection of integer k, we get $M_k = U_k S_k V_k^T$. Finally, we reduce the dimensions using formula $d^* = d^T V_k^T S_k^{-1}$, where d denotes the original feature vector of a specific snippet and d^* denotes a new one after dimensions reduction.

3.2 Category-Names Extraction and Snippets Distribution

In VSM, the distance between two feature vectors is calculated by *cosine distance*. Here, we use this distance to find out the most representative name for each category. Since category names may be not only words but also phrases, we reconstruct a new phrase-snippet matrix R with the dimension $m \times (r + n)$, where r denotes the total number of the phrases in the whole corpus. In subsection 3.1, after SVD, each column vector of matrix U_k denotes the abstract concept for a snippet. We can calculate the *cosine distance* between each abstract concept vector and each vector in matrix R. Then we can get the distance matrix $D = U_k^T R$. For each row vector, we find the maximum score whose corresponding term or phrase can be regarded as a candidate category name. Then regarding the candidate names as snippets, we construct another term-snippet matrix S with scores as its elements. After length normalization, we can get the matrix containing similarities between candidate names by formula $S^T S$. For each row of this matrix, only the name with the highest score will be retained.

After the category-names extraction, we classify all the snippets into the extracted categories. This work is done by *Snippets Distributor* block as shown in Figure 1. The concrete process is described as follows: we define a matrix P in which each category label is represented as a column vector. Then we can get the clustering matrix $C = P^T R$ whose element c_{ij} denotes the similarity between the jth snippet and the ith category. If similarity c_{ij} in C exceeds the predefined distribution threshold value, the jth snippet will be classified into the ith category. Moreover, we regard "category" as a field and store the names and the snippets of each category into the database.

3.3   Interesting Blocks Extraction

Since one blog page maybe contains not only the information blocks users are interested in but also the ones about other topics, we should find out the Information Content Blocks of Interests (ICBIs) before extraction. We first use Tidy[2] to transform the blog pages, which are retrieved according to the URLs stored in the database, into well-formed documents. In [2], Nanno et al. gave one of the blog's important features as follows:

Definition 1. *A page is judged to be a blog iff a sequence of entries that are articles for a day can be extracted from the page.*

Based on Definition 1, in this paper, we use the dated entries to determine the information block boundaries rather than to judge a blog page. We use regular expressions [8] to describe the possible date formats. For instance, "22-May-2007" and "22 May 2007" can be described by regular expression "$(\d+)\.?-?\s?([\d\.\s-]+)\.?-?\s?(\d\{4\})$". Then we use these regular expression to identify the dates in original Web documents. We also transform the matched date into an uniform date format, regard "date" as a field and store the date for each snippet into the database. By large amount of observations, we find that the date's outside but nearest tag <DIV> and tag </DIV> and the characters between these two tags actually form the information blocks. Moreover, the blocks which contain the snippets searched and stored in the database are ICBIs. Here, we propose an approach for judging whether a block contains a snippet or not. We have following definition:

Definition 2. *String $\mathcal{E}$ is a snippet and string $\mathcal{B}$ is an information block. Function $ws(s)$ can return the set containing all the words appearing in string s. $\mathcal{E}$ is contained by $\mathcal{B}$ iff $coverage(\mathcal{E}, \mathcal{B}) \geqslant \rho$, where $coverage(\mathcal{E}, \mathcal{B}) = \frac{|ws(\mathcal{E}) \cap ws(\mathcal{B})|}{|ws(\mathcal{E})|}$.*

Based on Definition 2, given a specific snippet, we calculate the *coverage* for each block and choose the one with the highest *coverage* as the ICBI. The above task is fulfilled by *ICBIs Extractor* block in Figure 1.

4   Knowledge Generation

We represent knowledge by OWL statements written in 3-tuples $\mathcal{K}=(subject, predicate, object)$. Extracting knowledge is actually a process of extracting subjects, predicates and objects. Every time when we execute an extraction, we only focus on one paragraph in an ICBI. Firstly, we do Chinese lexical analysis using ICTCLAS [7], including word segmentation and part-of-speech tagging. Then we

---

[2] http://www.w3.org/People/Raggett/tidy

can get all the words and their parts of speech. Secondly, we invoke ICTPROP[3] to fulfill the parsing task and finally identify the subject, the predicate (core verb), and the object for each sentence. According to the core verbs, we classify the sentences into three types as follows:

1. "做(Do)" statements (ex. *CPCC promulgated a bulletin yesterday.*);
2. "是(Be)" statements (ex. *The ID number of CPCC stock is 600028.*);
3. "成为(Become)" statements (ex. *The price of CPCC becomes 12.33.*).

For the "做(Do)" case, we regard the core verb itself as a property name; for the "是(Be)" case and the " 成为(Become)" case, we ignore the core verb and regard its subject as a property name. Since Web resources are formatted in diverse ways for human viewing, we should map the original property names to the ones predefined in domain ontologies. *Names Mapper* block will do this work. Based on HowNet [9], a *Names Dictionary (ND)* is maintained to store the original names and the predefined names. We invoke the *ND* to fulfill the *name mapping* task. By analyzing the domain and the range based on domain ontologies, the subject and the object of the property corresponding to the identified verb can be automatically tagged. Then individuals are generated based on the identified 3-tuples. We regard a specific individual's class name plus a random integer as the individual's URI to guarantee that the URIs are unique. For each statement, we assign an ID number to it after reification. Finally, we store the ID number, the reified statement and the original paragraph into the database. The above task is fulfilled by *Knowledge Generator* block as shown in Figure 1.

5 Rules Construction and Knowledge Reasoning

In our system, we invoke KAON2[4] to fulfill the reasoning task. Besides the constraints predefined with OWL in domain ontologies, sometimes users maybe want to add new rules to the KB. We classify these rules into two types: common sense rules and personalized rules. The common sense rules describe some widely accepted facts. Suppose a company issues a stock and the annual report of the company tells that the retained profits exceed 10% of the stock price. Based on the above statements, this stock is regarded as a fine stock. This fact can be described as the following rule:

(**true**, [*is_a*, s, fine_stock]) :−
    (**true**, [*is_a*, s, stock]), (**true**, [*is_a*, c, company]), (**true**, [*is_a*, r, report]),
    (**true**, [*is_a*, p, retained_profits]), (**true**, [*issues*, c, s]), (**true**, [*release*, c, r]),
    (**true**, [*states*, r, p]), (**true**, [*has_value*, p, $1]), (**true**, [>, $1, 0.1])

Here, each pair of parentheses denotes a literal and each comma denotes *conjunction*. Between the parentheses, the pair of square brackets denotes a binary relationship. $X denotes a variable with the name X. The reserved words **true**

---
[3] http://www.nlp.org.cn/project
[4] http://kaon2.semanticweb.org

or **false** are used to assign values to literals. On the other hand, the personalized rules describe users' personalized opinions. Usually, these opinions are experiential. Suppose a user deems that if the current price of a stock is larger than 8 and the rate of the price fall is equal to or more than 10%, the stock is in trouble and should be sold. This opinion can be described as the following rule:

> (**true**, [is_a, s, trouble_stock) :−
>   (**true**, [is_a, s, stock]), (**true**, [$current\_price$, s, \$1]),
>   (**true**, [$price\_fall\_rate$, s, \$2]), (**true**, [>, \$1, 8]), (**flase**, [<, \$2, 0.1])

In KEROB, rules can be conveniently written by selecting candidate items on the user interface through *Rules Generator* block. Each rule has two parts: *head* and *body*, which are composed of several literals written in 3-tuples. When adding rules, users construct literals for *head* and *body* respectively by selecting items in drop-down lists, which contain the names of classes and properties used for tagging the documents in a specific category before.

6   Search and Results Presentation

We use the restrained NLP technique to deal with the users' queries. *Restrained* means users use text boxes to generate queries instead of using natural language. When a specific user asks a question, he first manually splits the question into three parts: subject, predicate and object. Then he fills in the boxes with the corresponding known parts and leave other boxes blank. We propose a method of constructing queries automatically. In our system, each query is composed of two sets: the condition set and the variable set. The condition set is composed of the literals describing the facts. On the other hand, the variable set is composed of the variables whose values users want to query. The pseudocode for constructing queries is shown in Algorithm 1. Then based on these two sets, we use KAON2 to

---

Algorithm 1. Query Construction Algorithm.

Input: A 3-tuples T containing the users' inputs in 3 boxes.
Output: The condition set and the variable set.

1. Create a 3-tuples $spo = (part_1, part_2, part_3)$;
2. Create a condition set *condition* and a variable set *variable*;
3. Map T to a new 3-tuples $S = (str_1, str_2, str_3)$ based on *ND*;
4. **for** (each string str in S){
     4.1. **if**(str_i is *null*){
          4.1.1. Create variable V_i; 4.1.2. Assign V_i to $part_i$; 4.1.3. $variable.add(V_i)$;}
     4.2. **else if**(str_i is the name of a class){
          4.2.1. Get the class C corresponding to str_i;
          4.2.2. Create *ins* as an instance of C; 4.2.3. Assign *ins* to $part_i$;
          4.2.4. Create a new literal $L =$(**true**, [is_a, ins, c]); 4.2.5. $condition.add(L)$;}
     4.3. **else if**(str_i is the name of a data type){
          4.3.1. Get the data type dt corresponding to str_i; 4.3.2. Assign dt to $part_i$;}
     4.4. **else if**(str_i is the name of a property){
          4.4.1. Get the property *pro* corresponding to str_i; 4.4.2. Assign *pro* to $part_i$;}
   }
5. Create a new literal $L =$(**true**, [$part_2$, $part_1$, $part_3$]);
6. $condition.add(L)$;

---

construct the queries and find out the values of the variables, actually the ones users expect to search. Finally, the search results will be displayed to users. In order to make the user further understand the results, we also display the whole 3-tuples statements and the original paragraphs from which these statements derive based on the ID numbers stored in the database.

7 Experiments

KEROB (**K**nowledge **E**xtracto**R** **O**f **B**logs) is carried out on a PC with an AMD Athlon 64 CPU with $1.9GHz$ and $1GB$ of RAM, using Java programming language. We retrieve 1509 blog documents by invoking Google AJAX Search APIs and taking "股票" (stock) as the inputted search phrase. After the pre-clustering, we obtain 14 categories (total 1433 documents) except category *other* which contains the unclustered documents (total 76 documents). We ask 30 users to generate 600 queries for all this categories. Focusing on these queries, we also ask eight domain experts to manually extract OWL statements and give the results. The domain ontologies used in our system are created by four persons within six months, including 200 classes, 265 properties and 1332 individuals. We evaluate the performance of our system by calculating *Recall*, *Precision* and $F_{measure}(\beta = 1)$, which are defined as follows:

$$Recall = \frac{|S_{total}| \bigcap |S_{DE}|}{|S_{DE}|} \times 100\%, \quad Precision = \frac{|S_{total}| \bigcap |S_{DE}|}{|S_{total}|} \times 100\%$$

$$F_{masure} = \frac{2 \times Recall \times Presion}{Recall + Precision} \times 100\% \quad (\beta = 1)$$

Here, S_{total} denotes the set of OWL statements extracted by our system and S_{DE} denotes the set of OWL statements extracted by domain experts manually. $|X|$ denotes the cardinality of set X. After running of our system, we deal with those queries again using the traditional IE technique. This technique dose not refer to any domain ontologies and directly stores the extracted data into XML

Table 1. Performance comparison

Category Name (In Chinese)	Without Ontologies			With Ontologies		
	Pre.	Rec.	F_1	Pre.	Rec.	F_1
知识股票入门	41.16	65.63	53.50	70.35	81.39	75.47
股票行情查询软件	76.25	75.35	75.80	91.45	81.74	86.32
股票网址大全	43.48	62.50	51.28	58.26	70.53	63.81
分析个股	65.05	77.84	70.87	77.08	83.67	80.24
股票论坛	67.87	75.12	71.31	84.94	83.08	84.00
上市公司基金	75.78	81.51	78.54	80.47	84.25	82.32
股票投资	56.82	77.46	65.55	73.55	92.23	81.84
提供股票证券	38.92	59.05	46.92	65.06	76.85	70.46
股票证券财经	60.24	77.77	67.90	72.81	85.28	78.55
股票机构	47.50	70.74	56.84	72.14	81.12	76.37
实时行情	65.06	66.79	65.91	68.03	73.49	70.66
东方财富	65.08	57.75	61.19	73.02	63.01	67.65
股票权证	58.33	66.96	62.35	64.39	71.43	67.73
股票银行	43.93	62.80	51.70	65.32	81.29	72.44

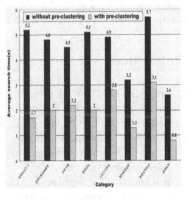

Fig. 2. Search time comparison

files or the database. Then we calculate *Recall*, *Precision* and $F_{measure}(\beta = 1)$ of extracting information from the retrieved blogs without domain ontologies and with domain ontologies respectively and list $F_{measure}$ in Table 1. The result shows that the number of correct statements extracted with domain ontologies is larger. Especially, for some category, this number for "with the pre-clustering" case is improved to be nearly twice as many as the one for "without the pre-clustering" case. All in all, *Recall*, *Precision* and $F_{measure}(\beta = 1)$ are all improved when domain ontologies are invoked.

We also compare the efficiency of querying the KB generated without the pre-clustering with that of querying the KB generated with the pre-clustering. We first select eight categories in which the extracted statements are more than those in other categories. Within the process of querying the KB, we record the average searching time for each category, as shown in Figure 2. From this figure, we can see that the average searching times for querying with the pre-clustering are shorter than the ones for querying without the pre-clustering.

8   Conclusion

We propose a system for extracting knowledge from blogs in Chinese based on the pre-clustering. Moreover, a prototype of our system KEROB is implemented. Using this system, users can acquire the most relative knowledge which are stored in original blog pages or inferred by the reasoner. Our KB is timely updated in the background processing of KEROB. New rules can be created according to the users' specific needs. The experimental results indicate the superiority of our system. Since the performance of our system is restrained by the specific domain ontologies, this prototype focuses on only the stock market at present. Our long-term goal is to develop a more general knowledge search engine for blogs by importing other existing domain ontologies.

Acknowledgement

This work is supported by the NSFC (grant no. 60496321) and European Commission under the project EASTWEB (grant no. 111084).

References

1. Liu, J.S., Lee, C.Y.: Extracting Structured Subject Information from Digital Document Archives. In: Sugimoto, S., Hunter, J., Rauber, A., Morishima, A. (eds.) ICADL 2006. LNCS, vol. 4312, pp. 141–150. Springer, Heidelberg (2006)
2. Nanno, T., Suzuki, Y., Fujiki, T., Okumura, M.: Automatic collection and monitoring of Japanese weblogs. In: Proceedgings of the WWW 2004, ACM Press, New York (2004)
3. Kurashima, T., Tezuka, T., Tanaka, K.: Mining and visualizing local experiences from blog entries. In: Bressan, S., Küng, J., Wagner, R. (eds.) DEXA 2006. LNCS, vol. 4080, pp. 213–222. Springer, Heidelberg (2006)

4. Kleinberg, J.: Bursty and hierarchical structure in streams. In: Proceedings of the KDD 2002, pp. 91–101. ACM Press, New York (2002)
5. Nakatsuji, M., Miyoshi, Y., Otsuka, Y.: Innovation detection based on user-interest ontology of blog community. In: Cruz, I., Decker, S., Allemang, D., Preist, C., Schwabe, D., Mika, P., Uschold, M., Aroyo, L. (eds.) ISWC 2006. LNCS, vol. 4273, pp. 515–528. Springer, Heidelberg (2006)
6. Schürmann, K.B., Stoye, J.: Counting Suffix Arrays and Strings. In: Consens, M.P., Navarro, G. (eds.) SPIRE 2005. LNCS, vol. 3772, pp. 55–66. Springer, Heidelberg (2005)
7. Zhang, H.P., Yu, H.K., Xiong, D.Y., Liu, Q.: HHMM-based chinese lexical analyzer ICTCLAS. In: Proceedings of the 2nd SIGHAN Workshop, pp. 184–187 (July 2003)
8. Ilie, L., Shan, B., Yu, S.: Fast algorithms for extended regular expression matching and searching. In: Alt, H., Habib, M. (eds.) STACS 2003. LNCS, vol. 2607, pp. 179–190. Springer, Heidelberg (2003)
9. Gan, K.W., Wong, P.W.: Annotating information structures in Chinese texts using HowNet. In: Proceedings of the 2nd workshop on Chinese language processing, pp. 85–92 (2000)

Complex Information Management Using a Framework Supported by ECA Rules in XML

Bing Wu, Essam Mansour, and Kudakwashe Dube

School of Computing, Dublin Institute of Technology Kevin Street, Dublin 8, Ireland
`<firstname>.<surname>@dit.ie`

Abstract. It is every organization's desire to incorporate best practice into its enterprise. This incorporation gives rise to the need to maintain information that could be viewed as complex. Managing this complex information poses a major challenge to the area of information management. This paper presents a framework for the incorporation of best practice and subsequent management of the resulting complex information. The paper also presents an approach to supporting this framework by using the ECA rule paradigm with an XML-based language, called AIM, for specifying and querying best practice and the complex information.

Keywords: Information management, ECA rule paradigm, active database, XML, rule markup languages.

1 Introduction

Domains, such as patient care practice and securities trading order management, involve cross-domain activities that require the application of best practice, e.g., knowledge from research and experience, and also require constant monitoring of some vital signals or essential happenings within the domain. Computerizing best practice in the form of such *complex information*, as conceptualized here, could be seen to pose major challenges for information management. The first challenge is not only a matter of disseminating the best practice within application domains, but also a matter of ensuring the computer-interpretation for its specifications. The second challenge is to support the customization of best practice to suit a specific domain user and task. The third challenge is the incorporation of best practice into the computerized execution of tasks that uses the *complex information*, and keeping the execution history in order to review the evolution of the *complex information*. The fourth challenge is to facilitate the manipulation, query, and review by replay operation as well as sharing and distribution for this information.

This paper presents a novel approach and framework that incorporate application best practice into organizations' enterprise, and manage the complex information produced by such incorporation. Our approach and framework is restricted to applications that naturally take the form of reactive applications that monitor events of interest to domain users, and respond to changes in situations by issuing alerts, reminders, requests, and/or observations to domain users. In order to adapt to the

A. Paschke and Y. Biletskiy (Eds.): RuleML 2007, LNCS 4824, pp. 224–231, 2007.

needs of users, markets, and enterprise management, the decentralized management is considered as one of the key aspects of our approach and framework. The method adopted to realize the approach and framework aims at utilizing the generally available highly optimized and easily maintained technologies, such as the *Event-Condition-Action (ECA)* rule paradigm [1] and XML technologies, to provide a Web-enabled tool that assists domain experts and users in managing the *complex information*.

The rest of this paper is organised as follows: Section 2 discusses the problem of managing the complex information; Section 3 outlines related work; Section 4 presents our approach to modelling the complex information; Section 5 outlines the SEM framework for managing complex information through the three planes covering specification, execution and manipulation of the information; Section 6 explains the role played by ECA rule paradigm and XML in supporting the SEM framework; Section 7 presents a conceptual architecture for the proposed prototype system that implements the framework; and finally, Section 8 summarises and concludes this paper.

2 Problem of Managing Complex Information in Computerised Best Practice

Best practice exists in several forms, such as recommendations from recent research results, practical experience, expertise and new methods of doing things in a given domain. Incorporating the best practice into application domain tasks and activities results in the need to manage *complex information*, whose main components include: 1) The domain information or data items that is relevant and is therefore required to be monitored in the application of knowledge and best practice; 2) Knowledge from the best practice and experience that is applied in consideration to the user's preferences or situations; 3) A description or reference material associated with the specific area of focus; and 4) The history and experience arising from daily practice of using the best practice knowledge in the domain. Consider as example the stock market trading domain, in which the customer order is specified by a customer who intends to buy or sell securities at the prevailing market price. The customer orders have *conditions* and *constrains* that determine whether a transaction will occur and at what price. The customer order could be considered to be a body of *complex information* that consists of: 1) Information on particular stock items making up the order. This information is also a subject of monitoring; 2) Knowledge that represents constrains on the monitored stock items defined by the customer and the best practice that is used to continuously adjust the customer orders, as soon as stock items are changed, 3) The order history that shows the changes in the stock items and the constrains, over the life-cycle of the order. The order evolution history represents experience that assists in making decisions regarding future orders, 4) Descriptive information for the order, such as the order number and date, as well as any relevant didactic information that may be of help in educating stakeholders.

It is required to manage such information through the specification, execution and manipulation as one distinct entity using a high level and declarative method.

3   Related Work

It is generally recognised that best practice need to be incorporated into daily practice and further be provided through computer-based support mechanisms. Several works in different research areas, such as a) the active database; b) workflow and business process; c) AI and decision making, address the problem of incorporating best practice into application activities. Our work is distinguished from other research efforts as explained in the following paragraphs.

In the area of active database, several research efforts have utilized ECA rule paradigm [1] to combine best practice (business logic) and the database together into the DBMS, such as [2, 1]. According to this approach the best practice is represented as individual triggers, and the domain information is represented as relations (tables). Both, the best practice and the domain information are managed separately. This approach suffers from 1) the problems of unexpected interactions among the individual rules as the rule base size increases [2]; and 2) the gap between the end-user's needs of managing specific situations and the low-level representation using triggers. Our approach provides a high level and declarative manner to bridge this gap, and classifies the rules into sets.

In the area of workflow and business process, several research efforts, such as [3, 4], attempt to standardize the applications activities, according to the application best practice, as processes. This approach differs from ours in that it address the specification aspects with little or no support to query and manipulate all aspects of the information in a unified manner. However, our approach aims at automating the *complex information* that includes dynamic and static aspects within a unified framework that provides support to specify, query, and manipulate the dynamic and static aspects of the *complex information*.

Comparing with decision making and AI techniques, our work focuses only on assisting the decision-making process by issuing notifications, reminders, and/or observations regarding situation of interest to the domain user, or by executing predefined changes.

At a general level, existing works provide little or no support for comprehensive unified management of the information and processes derived from best practice in domain applications. Our approach facilitates the incorporation of best practice into application activities requiring constant monitoring, and provides decentralized management support for all information as one distinct entity produced from such incorporation.

Incorporating the ECA rule paradigm into XML has being addressed in several research areas, including *active XML, RuleML*, and *semantic Web*. In the area of active XML, the ECA rule paradigm [1] is incorporated into the XML to support active behavior over XML data, such as in [5, 6], which support the reactive applications at the level of rules and triggers only. The RuleML language aims at providing a standard rule language that is interoperability platform [7, 8]. Wanger in [7] classified the ECA rule paradigm as a subtype of the RuleML language. The RuleML language has been utilized to support semantic Web and business applications, such as in [9]. However, the RuleML language formalizes the application best practice as individual rules, not as a unified distinct entity.

4 An Approach for Modelling the Complex Information

Fig. 1 illustrates a model for the complex information by using a UML class diagram. The components of a complex information item for a given application domain are: a knowledge and action component that models aspects such as reactive behaviour, a domain information component that models monitored information, an evolution history component and a descriptive information component. These pieces of information could be classified and viewed as *active* and *passive* information.

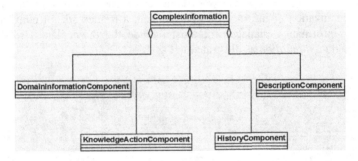

Fig. 1. The model for complex information

Here, the *active* information is modelled by the knowledge action component that determines the reaction that should be taken as response to changes in the monitored domain information. The knowledge action component could constitute reactive behaviour that could be practically represented as modularized sets of rules. The *passive* information is the evolution history as well as the domain and descriptive information, which are essentially of a factual nature.

The approach adopted in this study emphasizes the management of the *complex information* (CI) as one distinct entity as follows: The CI 1) has a general structure specified according to the best practice of an application; 2) deals with particular situations according to the user's preferences and interest; 3) is executed, as soon as a change of interest happens to the monitored information. The execution of the CI provides notification and suggestions to the user and might modify the knowledge action component; 4) could be manipulated and queried as a first class object, not only as row data; 5) is provided a distributed management that supports the remote users and distributed applications. The approach's unique features include: 1) managing the active and passive information of the CI as one object, not as a separated objects, such as in the active database the passive information represented as relations and the active information managed as individual triggers; 2) a high level declarative language for specifying, manipulating, and querying the CI; 3) utilizing an event-driven mechanism incorporated with XML technologies; 4) support for the decentralized management of the CI.

5 SEM Framework for Managing Complex Information

This Section presents a framework, called SEM, for incorporating best practice into the organization activities and managing the *complex information* produced from such incorporation. The SEM framework extends and enhances the framework developed by Dube in [10] for managing clinical guidelines. A key aspect of information management is the handling of knowledge acquired by one or many disparate sources in a way that optimizes access and use by all who have a share in that knowledge or a right to that knowledge. Information management in this work involves the administration of information, its uses and transmission and spans the entire process of defining, evaluating, protecting, and distributing data and information within an organization. It also includes the planning, budgeting, manipulating, and controlling of information throughout its life cycle. To be comprehensive enough to incorporate this conceptualization of information management, a framework for computer-based support for information management must include the three planes: *specification, execution* and *manipulation* as illustrated in Fig. 2.

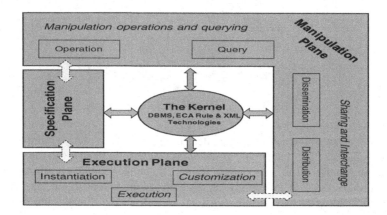

Fig. 2. SEM: The complex information management framework

The *Specification Plane* provides for the requirement of creating specifications that are essential to capture domain knowledge and information. It also involves specifications that are essential to enable the execution of instances. The *Execution Plane* focuses on meeting the requirement for allowing the execution of the aspects of domain knowledge that is amenable to execution by using mainly an event-driven approach based on the ECA rule paradigm. However, other execution formalisms are also possible. The *Manipulation Plane* provides for the need to be able to manage domain knowledge and information through: performing manipulation operations; issuing various types and forms of queries; enabling the sharing by allowing dissemination and interchange of knowledge and information; and providing for all these within distributed contexts. The dotted white arrows in Fig. 2 illustrate the fact that these three planes also interact in a dynamic fashion and offer services to one another.

6 The Role of ECA Rule Paradigm and XML in Supporting SEM Framework

Realizing SEM as a unified framework for managing the *complex information* (CI), in the reactive applications requires technologies that: 1) can be seamlessly integrated and easily incorporated, 2) support the monitoring process and distributed management. The ECA rule paradigm incorporated into XML technologies are utilized here to support the management of the CI. Furthermore, the modern DBMSs, such as Oracle and DB2, provide support for ECA rule paradigm and XML technologies. Consequently, the event-driven and XML technologies are adopted to play a crucial role in providing the basis for the management of the CI. The benefits of this method include: the flexibility of managing the CI as one unit, and the easy integration of the complex information management system into other systems. This method facilitates the development of the proposed approach for modeling the CI, a management language and a decentralized system for the complex information. Based on this method a language, called AIM [11], is developed such that: 1) it is an XML-based language and enjoys the general benefits of XML, such as parser reuse, incorporation into Web services, query generation; 2) It has an ECA- and XML-based specification component language, called AIM-SL, which formalizes the application best practice into interpretable format; 3) it has a high level XQuery-based component language, AIMQL, that provides support to manipulate, query, and has an extension for replaying the CI. The AIM language provides a CI model, which is developed as an active temporal XML document. This document includes: 1) the knowledge and action component represented as ECA rules; 2) the history component represented using temporal XML mechanism; 3) the domain information and descriptive components represented using conventional XML. The service-oriented architecture incorporated into the modern DBMS is utilized to develop a system, called AIMS. AIMS manages the complex information in distributed environment, as outlined in Section 7.

The SEM framework based on the use of ECA rules with specifications in XML offer a number of benefits. Firstly, the framework unifies all aspects of the management of information and fosters a holistic perspective to the management of the information and knowledge. Secondly, a high level of flexibility can be afforded by the ability to make on the fly specification and modifications during execution. Thirdly, the ECA rule paradigm in XML provides the best practical compromise for best practice knowledge incorporation into, and sharing among, information systems and integration with domain databases.

7 Conceptual Architecture for a Complex Information Management System

Fig. 3 illustrates the abstract and conceptual architecture for the proposed prototype system, called AIMS, that supports the *SEM framework* for managing complex information in terms of the three planes. The implementation of the *Manipulation Plane* is achieved through the query and manipulation management. The effect of the

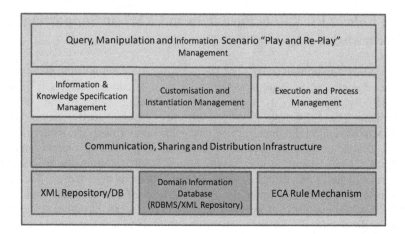

Fig. 3. Conceptual architecture for a system for supporting the SEM framework

manipulation plane will be reflected in the database as well as the ECA rule mechanism. The implementation of the *Specification Plane* is attained through the information and knowledge management, which allow users to specify knowledge using AIM [11].

The resulting specifications are stored in the database. The implementation for the *Execution Plane* is attained through the customization and instantiation, and the execution and process management functions as well as the ECA rule mechanism within the database or XML repository. The operations and queries of the manipulation plane should also be able to apply to information in both the execution plane and specification plane.

The SEM framework offers a comprehensive way to manage information such that the specifications and the execution process and results are enabled for the manipulation operations and queries within the same framework. The ECA rule paradigm as implemented within the context of databases or of XML repositories offer a practical way of incorporating best practice into routines used daily.

8 Summary and Conclusion

The incorporation of best practices into the computer-based tasks and activities of an organisation gives rise to the need to maintain information that could be viewed as complex. Managing this complex information poses a major challenge to the area of information management. This paper has presented the SEM framework for the incorporation of best practice and an approach to the subsequent management of the resulting complex information. Unlike others, the SEM framework does not focus only on one aspect of information management but incorporates planes for specification, execution and manipulation. The approach to supporting this framework is based on the use of the ECA rule paradigm within an XML-based language, called AIM, for specifying domain knowledge and information. Specifying the ECA rule paradigm in XML allows domain specific formal annotations and structuring of information and knowledge that is manageable and sharable across technologies and domains. This work demonstrates that RuleMLs provide a strong basis for realising the comprehensive SEM framework for knowledge and information management. Thus, every organization's desire to incorporate best practice into its enterprise would appear to be made easier.

References

1. Widom, J., Ceri, S.: Active Database Systems: Triggers and Rules For Advanced Database Processing. Morgan Kaufmann, San Francisco (1996)
2. Umeshwar, D., Blaustein, B.T., Buchmann, A.P., et al.: The HiPAC Project: Combining Active Databases and Timing Constraints. SIGMOD Record 17, 51–70 (1988)
3. Hong, H.-S., Lee, B.-S., Kim, K.-H., et al.: A Web-Based Transactional Workflow Monitoring System. In: WISE 2000. Proceedings of the First International Conference on Web Information Systems Engineering, IEEE Computer Society Press, Los Alamitos (2000)
4. Tsalgatidou, A., Athanasopoulos, G., Pantazoglou, M., et al.: Developing scientific workflows from heterogeneous services. SIGMOD Rec. 35(2), 22–28 (2006)
5. Abiteboul, S., Benjelloun, O., Manolescu, I., et al.: Active XML: A Data-Centric Perspective on Web Services. In: BDA 2002 (2002)
6. Bonifati, A., Braga, D., Campi, A., et al.: Active XQuery. In: ICDE 2002. Proceedings of the 19th International Conference on Data Engineering, San Jose (California) (2002)
7. Wagner, G., Antoniou, G., Tabet, S., et al.: The Abstract Syntax of RuleML - Towards a General Web Rule Language Framework. In: Web Intelligence (2004)
8. Wagner, G., Giurca, A., Lukichev, S.: A General Markup Framework for Integrity and Derivation Rules. In: Principles and Practices of Semantic Web Reasoning (2005)
9. Pontelli, E., Son, T.C., Baral, C.: A Framework for Composition and Inter-operation of Rules in the Semantic Web. In: RuleML 2006 (2006)
10. Dube, K.: A Generic Approach to Supporting the Management of Computerised Clinical Guidelines and Protocols. PhD Thesis, Dublin Institute of Technology (DIT) (2004)
11. Mansour, E., Dube, K., Wu, B.: AIM: An XML-Based Temporal and ECA Rule Language for Managing Complex Information. In: Paschke, A., Biletskiy, Y. (eds.) RuleML 2007. LNCS, vol. 4824, pp. 232–241. Springer, Heidelberg (2007)

AIM: An XML-Based ECA Rule Language for Supporting a Framework for Managing Complex Information

Essam Mansour, Kudakwashe Dube, and Bing Wu

School of Computing, Dublin Institute of Technology, Kevin Street, Dublin 8, Ireland
<firstname>.<surname>@dit.ie

Abstract. This paper presents an XML-based event-condition-action (ECA) rule language, AIM, for supporting the SEM framework and approach to the computer-based incorporation of best practice in daily work and the subsequent management of the resulting complex information. SEM framework provides knowledge and information management support in terms of three planes: the specification plane, the execution plane and the manipulation plane. AIM language is an assembly of declarative language modules for supporting the three planes of the SEM framework and envisages its use within the context of XML and databases.

Keywords: ECA rule paradigm, XML language, active database, clinical practice guidelines, information management.

1 Introduction

The formalization and computerization of best practice and its subsequence incorporation into an organisation's computer-based information systems results in *complex information* whose management poses a serious computing challenge. In intensive care applications, a medical patient plan is an example for the *complex information* that is produced by incorporating the healthcare best practice, clinical guidelines, into the disease management. The domain users are interested in incorporating the best practice and managing the *complex information* as one distinct entity- as it exists in the domain- and at high and declarative level. However, most of computerized approaches, which are utilized to incorporate best practice into application activities, focus only on incorporating the best practice at the level of: 1) individuals rules and triggers, such as in active database research [1, 2]; 2) processes that create a list of administrative actions based on the user's criteria, such as in workflow based approaches [3]; 3) decision making by utilizing a decision model or AI technique, such as in [4]. These approaches create gaps between the domain users and the computerized best practice. For example, it is difficult for doctors to review or modify the medical patient plan at the level of triggers or processes. Best practice needs to be specified using a suitable language based on an appropriate computational formalism. The language and computational formalisms used should be chosen specifically for its ability to support easy incorporation of best practice with domain

A. Paschke and Y. Biletskiy (Eds.): RuleML 2007, LNCS 4824, pp. 232–241, 2007.
© Springer-Verlag Berlin Heidelberg 2007

information databases as well as its ability to support computer-based execution mechanisms. The dynamic nature of best practice demands that specification languages and execution mechanisms should allow on-the-fly manipulation. Furthermore, the query and information retrieval facilities in information systems also need to be provided with respect to the formalised best practice. This work follows a generic, comprehensive and unified information management framework, called SEM [5], developed by the authors in order to address these challenges. The unified framework allows information to be managed comprehensively in three planes for the specification, execution and manipulation with each plane being able to be integrated with the other two planes.

This paper addresses the problem of providing a comprehensive language to support the unified framework, SEM, for managing complex information arising from the incorporation of best practice into computer-based applications. In this paper, a high level XML-based declarative language, called AIM, for supporting the SEM framework is presented. The rest of this paper is organised as follows: Section 2 outlines related work; Section 3 presents the AIM language requirements; Section 4 discusses the AIM language specification component, AIMSL; Section 5 presents the complex information model in AIM; Section 6 presents the AIM language query and manipulation component, AIMQL; Section 7 outlines the implementation progress and future work; Section 8 summarises and concludes this paper.

2 Related Work

Languages that support the incorporation of best practice into computer information systems has continued to attract a lot of research attention. The formalisation and computerisation of best practice and expert knowledge in the area of database administration has been achieved through using a XML-based rule language to specify business policies that govern user database access rights [6]. Thus, using computerised best practice and expertise, non-IT personnel could be allowed to perform functions that are normally performed by a database administrator. Further typical examples are found in the areas of business activity management (BAM) [7] as well as in automated e-business negotiation with emphasis on goals, policies, strategy and plans for decisions and actions [8] in which best practice is formalised and specified for the purpose of computerisation.

ECA-RuleML [9] constitutes work on an XML-based rule language that focuses on the logic programming framework for supporting specifications of event-condition-action (ECA) rules that are integrated with derivation rules and integrity constraints. In the area of active XML, the ECA rule paradigm [2] is incorporated into the XML to support active behaviour over XML data. Several active XML languages have been produced, such as Active XQuery [10], and An Event-Condition-Action language for XML [11]. These languages incorporate the best practice into the application activities at the level of separated rules and triggers, which makes difficulties to domain users to review or modify. The AIM language is unique in its provision of a unified framework that caters not only for creation of specifications but also for their execution and subsequent manipulation and querying within a comprehensive information management context.

3 AIM Language Requirements

The main requirements for AIM language is to support the *SEM framework* at the three planes that cover information and knowledge specification, execution, manipulation, querying and information scenario replays. These requirements are summarised as follows: 1) Language requirements for the Specification Plane: AIM language is required to support the specification process, in which the best practice is formally specified ; 2) Language requirements for the Execution Plane: with respect to a specific domain scenario, the best practice specification is customized and instantiated to produce the *complex information*, such as producing a medical patient plan for a specific patient using a specific clinical guidelines specification. The *complex information* contains a reactive behaviour that determines the correct reaction for certain situation, such as the medical patient plan gives medical recommendations when the patient temperature is changed. Hence, the best practice specification must be expressive enough to specify behaviour that can be executed within this plane; 3) Language requirement for the Manipulation Plane: The *complex information* is subject to the same manipulation operations, as the domain information, plus some special operations, such as terminate, enable, and disable. The manipulation operations could be included in the behaviour of the *complex information*, or issued by the domain users. The manipulation operations facilitate a) the propagation of the changes from the generic specification to the *complex information*; and b) the maintenance of the complex information. The *complex information* is also subject to the same queries, as the domain information, plus special query support, such as the replay function, which allow dynamic execution scenarios to be re-enacted for the user's review.

Thus, to satisfy the requirements of the SEM framework, AIM language must provide 1) a specification language, which we will call the *AIM Specification Language* (AIMSL); 2) a model for the complex information; 3) a manipulation and query language, which we will call the *AIM Query Language* (AIMQL).

4 AIMSL: The Specification Language for the Complex Information

This section presents the specification component of AIM language, which called AIMSL.

4.1 AIMSL Model and Distinguishing Features

The model of AIM Specification Language (AIMSL) follows the event-condition-action (ECA) rule paradigm. AIMSL model expresses the best practice as modularized sets of rules, which are classified according to functional objectives and scopes. Fig. 1 illustrates the XML Schema of the AIMSL model. In this schema model the best practice is formally specified as a protocol library, which consists of protocol specifications as well as specifications of global rules whose scope is the entire domain of discourse and one not associated with any protocol. As shown in Fig. 1, the individual protocols one made up of schedules and a set of protocols rules that not associated with any schedule. Each schedule is a set of rules that differs from an ordinary rule set in that it has an entry criteria and the fact that all rules in it are bound together by a common functional objective. Each rule in the specification is deemed to be an ECA rule, which is defined over some relevant domain information attributes. It should also be noted that protocol, schedule and rule element in the schema model has a set of attributes and that each element in the schema is made up of a sequence of a combination of attributes and other elements. Thus, the schema model allows ECA rules to be specified as either a memes of a set or a part of a protocol or a schedule element. It should be pointed out here that the protocol and the schedule are manageable as single units although they are effectively sets of rules. The header is a collection of pieces of release and didactic information. The release part provides information related to specific specification version. The didactic part provides literature related to the best practice; cites references to the source of the knowledge that is encapsulated in the AIMSL specification; and provides explanation.

The AIMSL schema is modularized to provide flexibility in modifying or enriching the AIMSL language to suit several application domains. For example, applications, which demand specific requirements for the condition part, could rU:\BPO\Lncs\4824\Pagination\Editing\48240236\48240236.doceplace the condition part with its own one.

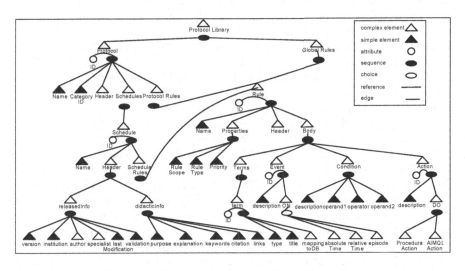

Fig. 1. The XML Schema for the AIMSL sub-language

4.2 ECA Rules and Temporal Features in AIMSL

In AIMSL model, the rule schema consists of elements (*name, properties, header, and body*) and the id attribute. The element *properties* determines the scope, type and priority of the rule. The scope specifies whether the rule is a global, protocol, or schedule rule. The rules, according to their event, are classified into to type static or dynamic rules. A static rule performs an action subject to the occurrence of a time event. A dynamic rule is a rule, whose event is a non-temporal event. The body consists of elements (*Terms, Event, Condition, and Action*). General terms are used in specifying the rule event, condition, and action. The element *Terms* specifies a general term and maps it into particular data items according to the utilized database schema. The term might be of type event or element. Consider as example, Rule 1: on two days after patient admission, order the blood test. AIMSL rule specification for Rule 1 will contains term, whose title and type are patient admission and event, respectively. The type event means this term will be used in the specification of the event of Rule 1. If the term is of type event, it will be mapped into database operation(s). If the term is of type element, it will be mapped into database attribute.

The event part of the AIMSL rules might be of type episode; absolute time; or relative time. The type episode means a domain event, such as "on patient admission", or "on receiving ACR test result". The event of type absolute time or relative time is a temporal event. The relative time event is a time event happening once or repeatedly and its time is related to a term of type event. Consider as example for once off event, on day 2 of patient admission and 2 hours before the completion time. Consider as example for repetitive event, every 3 days after patient admission for 10 times, or every 10 hours before the operation time.   The relativeTime element has a complex type composed of a choice between two elements, namely, onceOff and every. The absolute time is such as first of June 2008.

The condition part is expressing a simple condition consisting of two operands and an operator. The term of type element could be used as an operand to express a condition, such as ACR test result is greater than 25. The action part might be a procedural action, such as sending email, or an AIMQL action for manipulating or querying the complex information. More complicated conditions and composite events are considered as part of the future work.

```
-<protocol id="ProID-MAS">
    <name> microalbuminuria screening (MAS) protocol </name>
    <categoryID>CID316</categoryID>
    +<header>
    -<Schedules>
        -<schedule id="SIDMAS">
            <name>Basic MAS</name>
            +<header>
            -<scheduleRules>
                +<rule id="MAS1">
                -<rule id="MAS2">
                    <name>Rule 2 of basic MAS</name>
                    -<properties>
                        <ruleScope>Schedule</ruleScope>
                        <ruleType>Dynamic</ruleType>
                        <priority>0</priority>
                    </properties>
                    + <header>

                    - <body>
                        -<Terms >
                            <term id="E2.1">
                                <type>event</type>
                                <title>ACR test Result Received</ title >
                                +<mappingToDB>
                            </term>
                            <term id="E2.2">
                                <type>element</type>
                                <title>ACR test result value</ title >
                                +<mappingToDB>
                            </term>
                        </ Terms >
                        - <event id="E1R2">
                            <on>
                                <relativeTime>
                                    <onceOff>
                                        <granularity>hours</exsd:granularity>
                                        <timeLength >2</exsd:amount>
                                        <of>
                                            <term id=" E2.1">ACR test Result Received</term>
                                        </of>
                                    </onceOff></relativeTime></on>
                                </exsd:event>
                        +<condition id="ID36">
                        - <action id="AID36">
                        - <do>
                            -<AIMQLAction>
                                - <add>
                                    +<rule id="MAS3">
                                    +<rule id="MAS4">
                                </add>
                            </AIM-QLAction>
                        </do> </action> </body></rule></scheduleRules></schedule>
                    </Schedules></protocol>
```

Fig. 2. The MAS protocol specified using AIMSL

4.3 Example

Fig. 2 illustrates an example for an AIMSL specification of a simplified version of the microalbuminuria screening (MAS) protocol, which has a schedule containing two rules, MAS1 and MAS 2, as shown below. MAS2 defines a set of clinical recommendation that should happen two hours after the result of the required test in MAS1 is received. As shown in Fig. 2, the action of the rule MAS2 adds two rules to the specification, MAS3 and MAS4. The both rules are similar to the rule MAS1, but they fire on day 6 and day 38 of the patient admission, respectively.

Rule MAS1: **ON** day 2 of the patient admission,
       DO order the test albumin creatine ratio (ACR).

Rule MAS2: **ON** 2 hours after receiving the result of test ACR
       IF the ACR result is greater than 25
       DO order ACR test twice on days
       number 6 and 38 of the patient admission

5   The Complex Information Model in AIM

This section outlines a model for the *complex information* (CI). The CI model is presented in terms of the life-cycle, and structure.

5.1   Life-Cycle of the Complex Information

During the life-cycle of CI, the CI goes through state transitions, as shown in Fig. 3. These states are predefined and context-sensitive. The context-sensitive means that the CI's state is affected by changes in the domain information. When the CI is generated, it should be authorized to be registered or installed. In the registered state, all rules of the CI are installed in the system. In this state, no rule has fired yet. The CI moves to the "active" state once at least one rule is fired. The state "active" includes two sub-states, "waiting" and "executing". In the "waiting" state, at least one rule is fired and the other rules are waiting for events that are of interest

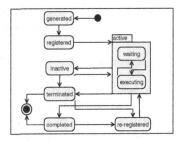

Fig. 3. A state diagram for the CI life-cycle.

to the CI. In the "executing" state, at least one rule is being executed. Once the rule execution completes, the CI returns to the "waiting" state. The CI might be transited from "active" state to "inactive", "terminated", or "completed" states. "inactive" state means that all the CI rules become disabled. The CI might be transited from "inactive" to "active" state. That means enabling the rules of the CI. "terminated" state means that all the CI rules removed from the system, but are not removed from the CI itself. When all the enabled rules in the CI are completed that means the CI is in the "completed" state. The "completed" state of the CI could be determined by a domain user, who is in charge of the CI. After the CI had become in the "completed" state, all the CI rules are removed from the system. It could be decided to re-register the CI again, after it had been terminated or completed.

5.2   Complex Information Schema

The CI consists of two main parts an active part and the passive part. The active part represents the reactive behaviour derived from the AIMSL specification. The passive part represents the descriptive information, state of the

CI and its evolution since it has been created. The passive part is subject to actions that log the execution history of the CI. Therefore, CI grows over time. CI is subject to dynamically changes in order to suit the current conditions and constrains of interest to the domain user. Fig. 4 illustrates the XML Schema for the CI model in AIM. As shown in Fig. 4, the active part of the CI is represented as rules. Each rule is coded as a trigger or several triggers. The trigger(s) are used to register the rule in the system. The rest of the XML Schema shows the passive part. The passive part of the CI is modelled as time-varying information. The model captures the valid times of the fact recorded under the CI. That is leading to temporal relations among the

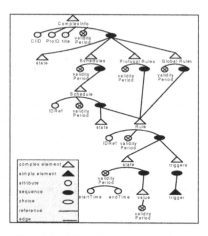

Fig. 4. The XML Schema for the Complex Information Model

CI and its components. On the other hand, the passive part shows the components of the CI, the validity period of their existence as a part of the CI, and their states. This model produces a temporal XML document, such as the document depicted in Fig. 5.

5.3 Example

A medical patient plan could be generated based on the specified protocol shown in Fig. 2. In the generation process, the rule body (terms,

event, condition and action) is used to generate a trigger, which could be encoded using SQL, SQL/XML, or XQuery triggering language. Choosing the triggering language depends on the type of the database used to store the domain information, whether it is a relational or XML database.

Assume 1) the medial patient plan is registered at time point 1; and 2) the result of ACR test

Fig. 5. A part of the patient plan on day number 4 of patient admission

is received on day 3 and its value is greater than 25. The action of MAS2 of the patient plan adds two new rules, MAS3 and MAS4, and then these changes are logged in the patient plan. Fig. 5 illustrates a portion of the patient plan on day 4. This portion has the history of the patient plan and its execution.

6 AIMQL: A High Level Query and Manipulation Language for the Complex Information

There is a need to move the complexity of manipulating and querying the best practice specified using AIMSL and its corresponding *complex information* from user/application code to a high level declarative language. AIMQL is a high level XQuery-based language provides facilities to perform manipulation operation, and advanced queries, such as replaying dynamic execution scenarios of the *complex information*.

6.1 Requirements

The main functional requirements of AIMQL are to assist in: 1) Manipulating the AIMSL specification and *complex information* (CI). The changes are made to AIMSL specification might be required to be propagated to the corresponding CI; and 2) Retrieving this information. This includes the ability to replay the CI or a specific part of it within specific time period. There are general functional requirements that should be also provided to AIMQL. These requirements are: 1) Declarativity, AIMQL should be declarative. It should be independent of any particular platform or query evaluation strategy. 2) Temporal Support, it should be able to record the history of executing the CI reactive behaviour and to query it. 3) XQuery-based, the AIMSL specification and CI are represented as XML document. Therefore, AIMQL should be based on XQuery. 4) Convenient for humans to read and write, this could be achieved using an XML-based graphical tool that assists in generating AIMQL query and browsing it. XML is easy to be generated using tools and easy to be converted to human readable format using a stylesheet language, such as XSL. Using XML in representing AIMQL provides a compatibility with AIMSL, and assists in managing the *complex information* remotely, using Web services.

6.2 Extensions to XQuery

Several extensions to XQuery are required in order to achieve the AIMQL requirements as following: 1) *Manipulation Operations*: AIMQL introduces seven manipulation operations (expressions). These expressions includes add, remove, modify, activate, deactivate, terminate and Fire. The AIMQL manipulation operations are distinguished in the sense that they not only potentially modify the AIMSL specification or CI, but also propagate the modification to the corresponding CI documents and modify the corresponding triggers created in the system. Furthermore, the manipulation expressions log the changes occurring to CI documents; 2) *Query Support*: AIMQL provides support to query AIMSL specification and CI document, as the domain information, plus special query capabilities, replay function and temporal query support for CI document, which is a temporal XML document.

AIMQL introduces a new functionality called replay. AIMQL replay query is a query that plays over again the history of the complex information to show in details the actions that cause changes on the complex information and how it evolved over time. Fig. 6 illustrates examples for AIMQL replay queries. These queries are presented here as patterns for AIMQL replay queries over medial patient plans, for short plans.

7   Implementation and Future Work

A combination of the ECA rule paradigm, XML, and database systems has been adopted as seamlessly integrated and easily incorporated technologies in the

implementation method for AIM language. The poof-of-concepts implementation of the first vision of AIM language is currently in progress. The AIM language is being implem ented using DB2, java, and XML technologies, s
uch as XQuery and Web services. The DB2 XML database is used to store specifications based on AIMSL schema. Several complex information documents could be generated from a specific AIMSL specification based on AIM model for the *complex information*. The main phase in generating a complex information document is the mapping of the rules into triggers. The temporal events, which discussed in Section 4.2, are not supported by the available DBMSs. We have extended the triggering mechanism of DBMSs to support the temporal events. The instantiation and execution of specifications is based on SQL

Replay Pattern 1:
Retrieve the history of the schedule no S1 of the plan no X, when the state of the rule no R of schedule S2 was ST. --AIMQL---------------------------------- REPLAY Conplex Information CI SHOW When, How, Why OF CI.schedule[@id = S1] CIS Where CI[@CIID = X] and CIS.overlaps (valid(CI.schedule[@id=S2]\rule[@id=R]\state[value = ST])) --Pattern Result---------------------------- This replay pattern returns the versions of Schedule no S1 of the complex information no X, such that the validity of the version overlaps the validity period of the state ST of rule R in schedule S2.
Replay Pattern 2:
Replay the plans of category no CAT, which was working through out the past Y days. --AIMQL-------------------------------- REPLAY Complex Information CI SHOW When, How, Why OF CI Where CI.cast("day") >= Y and CI.meets(NOW) and CI[@catID=CAT] --Pattern Result---------------------------- This replay pattern returns the versions of the plans of category CAT, whose validity period meets the current time, and whose age is greater than or equal Y days.

Fig. 6. AIMQL replay patterns.

trigger mechanism in DB2. However, our extension is a generic approach that could be applied to other DBMSs. The mapping between AIMSL and the SQL trigger language is being developed using Java, SQL in DB2.The AIMQL replay queries are transformed into our temporal XQuery language that is under implementation.

8   Summary and Conclusion

Most languages for the formalisation and specification of best practice are designed outside the context of a comprehensive management framework, sometimes leading to difficulties in managing the specified information or the integration of specifications with domain information. This paper has presented a high level XML-based declarative language, called AIM, for supporting the SEM framework. The SEM framework manages best practice through three planes covering specification, execution, manipulation and querying. AIM language aims at enabling the specification, behaviour execution, manipulation and querying of the *complex information* arising from handling computerised best practice. AIMSL, the specification component of AIM language, uses the ECA rule paradigm to allow the formalisation and specification of best practice to be incorporated into an event-driven mechanism using XML. The AIM model for *complex information* has been discussed. AIMQL is the manipulation and query component of AIM and is based on the XQuery language with promises to extend it with temporal facilities. Work on implementing the AIM language and evaluating it within the domain of clinical guideline management is currently on-going.

References

1. Umeshwar, D., Blaustein, B.T., Buchmann, A.P., et al.: The HiPAC Project: Combining Active Databases and Timing Constraints. SIGMOD Record 17, 51–70 (1988)

2. Widom, J., Ceri, S.: Active Database Systems: Triggers and Rules For Advanced Database Processing. Morgan Kaufmann, San Francisco (1996)
3. Tsalgatidou, A., Athanasopoulos, G., Pantazoglou, M., et al.: Developing scientific workflows from heterogeneous services. SIGMOD Rec. 35(2), 22–28 (2006)
4. Azwina, M.Y.: An on-line purchasing and decision support system for distributed retail chain stores. In: Proceedings of the 6th international conference on Electronic commerce, ACM Press, Delft, The Netherlands (2004)
5. Wu, B., Mansour, E., Dube, K.: Complex Information Management Using a Framework Supported by ECA Rules in XML. In: RuleML 2007, submitted to: International RuleML Symposium on Rule Interchange and Applications, Orlando, Florida (2007)
6. Gupta, A., Bhide, M., Mohania, M.: Towards Bringing Database Management Task in the Realm of IT non-Experts. In: ICDE 2003. Proceedings of the 19th International Conference on Data Engineering, IEEE Computer Society Press, Los Alamitos (2003)
7. Jeng, J.-J., Chang, H., Chung, J.-Y.: A Policy Framework for Business Activity Management. In: CEC 2003. IEEE International Conference on E-Commerce Technology, IEEE Computer Society Press, Los Alamitos (2003)
8. Li, H., Su, S.Y.W., Lam, H.: On Automated e-Business Negotiations: Goal, Policy, Strategy, and Plans of Decision and Action. Journal of Organizational Computing and Electronic Commerce 16(1), 1–29 (2006)
9. Paschke, A.: ECA-RuleML/ECA-LP: A Homogeneous Event-Condition-Action Logic Programming Language. In: RuleML 2006. Int. Conf. of Rule Markup Languages, Athens, Georgia, USA (2006)
10. Bonifati, A., Braga, D., Campi, A.: Active XQuery. In: ICDE 2002. Proceedings of the 19th International Conference on Data Engineering, San Jose (California) (2002)
11. Bailey, J., Poulovassilis, A., Wood, P.T.: An Event-Condition-Action Language for XML. In: www 2002. The 12th International World Wide Web Conference, Hawaii (2002)

Author Index

Lecture Notes in Computer Science

Sublibrary 2: Programming and Software Engineering

For information about Vols. 1– 4157
please contact your bookseller or Springer